ATTITUDE OR LATITUDE?

Attitude or Latitude?

Australian aviation safety

GRAHAM R. BRAITHWAITE
University of New South Wales

Routledge
Taylor & Francis Group

LONDON AND NEW YORK

First published 2001 by Ashgate Publishing

2 Park Square, Milton Park, Abingdon, Oxon OX14 4RN
711 Third Avenue, New York, NY 10017, USA

Routledge is an imprint of the Taylor & Francis Group, an informa business

First issued in paperback 2016

British Library Cataloguing in Publication Data
Braithwaite, Graham R.
   Attitude or latitude : Australian aviation safety. -
(Studies in aviation psychology and human factors)
   1.Aeronautics - Safety measures - Australia 2.Aircraft
accidents - Australia Statistics 3.Aircraft accidents -
Human factors - Australia
   I.Title
   363.1'24'0994

Library of Congress Control Number: 2001089099

Transferred to Digital Printing in 2014

ISBN 978-0-7546-1709-9 (hbk)
ISBN 978-1-138-26389-5 (pbk)

Contents

List of Figures

List of Tables

Foreword

I was delighted when Dr Graham Braithwaite asked me to write the Foreword to his book as I have been associated with Graham throughout the project. In 1993 when I was Manager Flight Safety at Qantas, I was forwarded a letter introducing a Loughborough University student who was studying aviation safety in Australia. I was somewhat puzzled by his request to see me as I was firmly convinced that aviation was safer in Australian than anywhere else - so why bother? It did not take me long to find out that Graham's aim was to establish whether this was true and, if it was, what could be learned from it. The results of his research would have a profound effect on my thinking about safety systems and their effectiveness.

The recent recognition of the role of human error in arguably 100% of all aircraft accidents has helped to create a new focus in safety. The system safety approach affords a better understanding of how aviation and other similar industries work. Yet throughout the evolution of safety investigations, there has been a tendency to analyse what went wrong, rather than what went right. To learn from a study of a good practices is the driving force of this book and is preferable to being caught up in the management by hindsight that often follows a catastrophe.

My next position in the Aviation industry was Deputy Chairman of AirServices Australia. AirServices is the Air Traffic service provider and was formed out of a split of the previous Civil Aviation Authority. The new organisation had no safety department and it was part of my task to ensure that this was established.

As part of his research Graham had made an astonishing discovery. He had put out a questionnaire to the industry asking which factors, from a given list, made the greatest contribution to the good Australian accident record. The response from the pilots indicated that 'Luck' was the fifth greatest factor. More worrying was that Air Traffic Controllers gave 'Luck' as the first factor! I obviously had some work to do.

During my time in Qantas management I assisted on a Committee to set up an Aviation Degree programme at the University of New South Wales, and shortly after started teaching Flight Safety as part of the degree.

I received a Research Grant to investigate Crew Resource Management for the *ab initio* pilot. This investigation required a Research Assistant. Fortunately by this time Graham had completed his Doctorate and took up this position. Perhaps my major contribution to Australian flight safety was getting Graham to come to Australia!

In my view the greatest threat to Australian flight safety has been and to a certain extent still is, complacency. Because we had not had a major public transport aircraft accident, it was extremely difficult to persuade the corporate world that money and effort still needed to be expended. The book addresses this issue and, as is so often the case, recent events have given everyone a timely reminder of the threat that complacency poses. This book could not have come at a better time and is recommended for all those in the industry - there is much to learn from it.

John Faulkner
Adjunct Associate Professor
Department of Aviation,
University of New South Wales

Preface

As a graduating student in 1993, I was determined that I wanted to work in aviation safety. This was largely a result of the inspiration provided by Dr Bob Caves of Loughborough University. Bob convinced me, and doubtless many others on the Transport Management and Planning course, that there was still a great deal to be done. Bob is one of those special people who buzzes with enthusiasm and it was his idea to look at Australian aviation safety.

The bulk of the material contained within this book is the product of doctoral research conducted at Loughborough and also based at RMIT and the University of New South Wales. It represents an attempt to contribute to the pool of knowledge across a number of disciplines both within aviation and in the wider community. Using both primary and secondary data and a structured methodology, the research was designed, not to prove a particular point, but rather to promote a system level understanding of the factors behind aviation safety.

The book is designed to provide a reference text which can be used both by academia and industry alike. Some of the principles explored are unique to aviation, but conversely there are many generic principles that are of use to other transport modes and indeed in any high-technology system. The structure of the case study aims to be accessible to both the aviation expert and non-expert alike.

The reader may not agree with every statement or interpretation and contrary views are welcome. The aim of the book is to stimulate, not dictate. Debate in safety is healthy - if there was only one right answer, life would be much simpler. If this book makes the reader think differently, then it has accomplished its primary objective.

My research was supported by many people and groups around the world. All deserve sincere thanks, not just for helping me, but also for being part of the continuing efforts in aviation safety. The support of the Transport Studies group at Loughborough University was unwavering throughout. In particular my research director, Dr David Gillingwater and supervisors Dr

Bob Caves and Professor Norman Ashford. During the dark days of trying to convince engineers that social science has a major role to play in aviation safety, the team provided inspiration and guidance. Thanks also to my colleagues at RMIT and the University of New South Wales, especially A/Prof. John Clements and Prof. Jason Middleton.

Similarly, the generous support of Rotary International Foundation is gratefully acknowledged. As an Ambassadorial Scholar for the Northwest Region of England, I was able to travel to the Royal Melbourne Institute of Technology to conduct a year's field research. In particular, the amazing *service above self* demonstrated by Desmond Carroll, Alan Williamson and Bruce Cameron made every difference to my work.

As for the aviation industry, it is impossible to name every pilot, air traffic controller, engineer, investigator, flight attendant or manager that helped me in this project. For the contribution they made to my research and continue to make to the safety of the aviation system every day, I offer my great thanks.

In particular, I wish to mention the contribution of a number of individuals. At Ansett Australia, Captain Max McGregor was always happy to help, as were Ken Lewis, Mike Innes, Michael Kemmis and Bev Maunsell at Qantas. Dr. Rob Lee, whom many will know as the Director of the former Bureau of Air Safety Investigation, has worked tirelessly to advance understanding of aviation human factors and always spared the time to help me out. Others who gave up valuable time with enthusiasm include Captain Peter Budd, Captain Neville Dickson, John Guselli, Brent Hayward, Dr Alan Hobbs, Captain Lionel Jenkins, Dr John Lane, Dr Claire Marrison, Paul Mayes, Dr Ashleigh Merritt, Group Captain Lindsay Naylor, Wing Commander Angela Rhodes, Paul Russell, Geoff Smith and Don Whitford.

Captain John Faulkner, former Manager, Flight Safety for Qantas and now Adjunct Associate Professor at the University of New South Wales deserves a special mention. John's experience, guidance and friendship has been invaluable throughout.

In Europe, Captain Stuart Grieve has become a great friend and mentor. Thanks also goes to Captain Heino Caesar, Air Commodore Richard Peacock-Edwards, Captain Neil Johnston, Captain George Robertson, Captain Colin Sharples, Captain Mike Wood, Frank Taylor and the AAIB's Ken Smart. The team at Ashgate also deserve sincere thanks for their guidance and efforts; John Hindley, Caroline Court and Pauline Beavers.

I am extremely sad that the man who has inspired me the most will never be able to read my work. The late Dr Roger Green will be known to most people working in aviation safety and human factors. To all that had the privilege to meet Roger, his enthusiasm for the subject was instantly inspiring. I once saw him present to a Royal Aeronautical Society meeting where he managed to transform a normally very sedate audience into a room bursting with excitement. His work continues through the many that he has motivated.

Finally, thanks to those closest to me who have supported me throughout. To my parents; Roy and Lyn, and my sister, Kerynne (the original Dr. Braithwaite); to my good friends Andrew Chilcott, Ben Thomas, Matt Greaves, Phil Dixon, Shalend Scott and Louis Balfour.

Last, but certainly not least, my eternal thanks to Lucy who has been the greatest tower of strength of all. Without you, I would have never finished writing this book.

GRAHAM R. BRAITHWAITE
Sydney, May 2001

1 Australia's Safety Record

Why use Australian Aviation as a Case Study?

The International Society of Air Safety Investigators' Code of Ethics
(ISASI, 1983) states that it is the responsibility of air safety investigators to
'...apply facts and analyses to develop findings and recommendations that
will improve aviation safety.' As such, one of their guiding principles is to
'...identify from an investigation those cause and effect relationships about
which something can be done reasonably to prevent similar accidents.'

Although ISASI states within its title that its concern is *air safety
investigation*, the terminology used within the Code of Ethics lends itself
particularly towards *accident* investigation and not *safety* investigation.
This is not unusual and is a good example of where the emphasis in safety
has traditionally lain.

Aviation, like most human activities has acquired learning through
its own mistakes. Many lives have been lost, in accidents that would now
seem unthinkable, in a pioneering sacrifice to progress. Accident investigators have become incredibly skillful in deducing *what went wrong* from the
remains of smouldering wreckage, but the price of progress using this
method is incredibly high. The shift to proactive safety has been a slow one
and one that is far from complete. The ability to learn from best practice to
avoid mishaps in the first place is the most effective goal for safety research
and an aim of this book.

It is widely suggested that Australian airlines exhibit a level of safety that is well above average. The statistics presented in this chapter support
that assertion for commercial jet regular public transport (RPT) operations
and if it could be proven that there are controllable variables behind the situation, then there is considerable merit in highlighting them and their relative contribution.

Industrial Psychologist, James Reason (1993) writes, 'Should we
not be studying what makes organisations relatively safe rather than
focussing upon their moments of unsafety? Would it not be a good idea to
identify the safest carrier, the most reliable maintainer and the best ATC sys-

tem and then try to find out what makes them good and whether or not these ingredients could be bottled and handed on?'

Aviation is a good example of a complex, socio-technical system, which, whilst arguably unique in some ways, is actually remarkably similar to other industrial systems both within transportation and outside. The high priority placed on aviation safety provides a number of excellent lessons for other industries without the high price of making their own mistakes.

This book is not concerned with either proving Australia to be the safest place to fly, or indeed its airlines to be the safest in the world - such an exercise is counterproductive and all but impossible. The main objective is to establish what is good about the Australian system so that it can be maintained and passed on to others in aviation and elsewhere. Such proactive safety is a more skilled and complex science than *management by hindsight* and therefore tends to have been overlooked in many areas, yet its value cannot be denied.

Geographical areas such as Europe or North America have excellent records for airline safety and benefit (statistically) from the stability of a large sample of take-offs and landings, especially in the case of the USA, which accounts for approximately half of the world's aviation activity. However, these regions make poor safety case-studies for a number of reasons including:

- Europe is covered by multiple regulatory / ATC authorities
- North America is covered by two separate regulatory / ATC authorities
- Europe represents an area of many diverse national cultures / languages
- Europe represents a wide variety of economic strengths
- Both represent a very large number of different airlines
- Both areas have a comparatively high turnover of airlines
- The majority of previous aviation safety work has focussed there.

Australia benefits from being a self contained continent of similar physical area to both the USA and Europe. The airlines contained within are relatively stable and all controlled under one regulatory and one air traffic control authority. There is a single common language, one Federal Government and there are relatively few RPT carriers. While the consideration of an entire aviation system involves the consideration of a very large number of variables, using Australia means some variables can be limited.

Is Australia's Record for Airline Safety Good?

Before any research can commence which aims to examine how a good safety record has been achieved, it is logical that *good* can be defined. Immediately, any hopes of a simple answer are dashed by the relative nature of the term *good*. If a comparison were made between the hull loss rate of any modern day airline to that of, say, the early US Post Office aircraft then the airline will appear to be excellent. If a similar comparison was made between the hull loss rate of even modern aircraft to the catastrophic loss of nuclear reactors then suddenly the loss rate may appear to be quite poor.

There are many ways to compare the safety of airlines and at its most basic, this involves a direct comparison of accidents over a period of time. Such a simplistic measure could be highly misleading and of little practical value. Airlines vary in size, history, aircraft types, operating areas and so on. It would be impossible to find two operations with enough similarities to make a comparison feasible and so the most simplistic tool available comes in terms of the accident rate. For two or more operations, one may consider the number of accidents per million departures or sectors.

In 1984, *Flight International* compared the safety records of leading air transport countries (Taylor, 1988). Safety was expressed in terms of fatal events per million departures as shown in Table 1.1.

Whilst the ten year period used is a relatively short sampling frame in terms of statistical validity, it also covers a number of technological advances such as the widespread introduction of GPWS and TCAS equipment. Key events may be seen to hold loud statistical noise. For example, the statistic presented for Turkey places it second highest in terms of fatal events and highest (by a long way) in terms of fatalities per 1 million flights.

In 1974, a THY Turkish Airlines DC10 suffered an explosive decompression near Ermenonville, France when its rear cargo-door blew out. Catastrophic damage to all of the aircraft's control systems, which run together near to the tail, led to a total loss of handling and the loss of 346 souls. Other similar incidents involving the DC10 aircraft demonstrated a poorly designed locking mechanism on the aircraft, which could easily be forced into a fail-unsafe position. The accident suffered by THY could have potentially befallen any DC10 operator and so could have easily skewed any number of the countries' records in Table 1.1.

Table 1.1   Aviation safety record for major nations

	fatal events per million flights	fatalities per million flights
Argentina	3.493	75.982
Australia	0.328	0.656
Belgium	1.938	203.488
Brazil	4.351	148.295
Canada	2.906	24.497
Colombia	27.464	509.269
Egypt	13.423	362.416
France	1.232	2.156
West Germany	1.298	27.261
India	5.512	333.858
Italy	1.772	114.311
Japan	0.642	13.262
Netherlands	2.523	383.515
Scandinavia	0.455	8.417
Turkey	17.369	1717.121
UK	1.279	56.763
USA	1.179	36.593
Venezuela	4.082	161.632

Source: Adapted from Taylor, 1988.

Using the measures from Table 1.1, *Flight International* produced an ranking of relative safety records, which is at least a starting point for even the most cynical researcher. As shown in Table 1.2, even this very basic analysis indicates a relationship between safety and economics as all of the top ten nations may be considered to be *first world* and the only developing countries appear in the bottom third. However, the need for caution is also apparent. In 1985, a Japanese B747 crashed as a result of structural failure with the loss of 520 souls. The third place held by Japan in *Flight International's* analysis would have dropped dramatically if the table were extended to cover 11 years.

Table 1.2   **Prime ranking 1973-1984 (based on fatal events per million flights)**

1	Australia	10	Netherlands
2	Scandinavia	11	Canada
3	Japan	12	Argentina
4	USA	13	Venezuela
5=	France	14	Brazil
5=	UK	15	India
5=	West Germany	16	Egypt
8	Italy	17	Turkey
9	Belgium	18	Colombia

Source: Adapted from Taylor, 1988.

The temptation is therefore to regionalise statistics in an attempt to smooth out statistical noise and form a larger base of data from which to work. The problem comes in attempting to define regions in a way that is both fair and meaningful. The simplest way is to use the geographic continents and both the International Air Transport Association (IATA) and Boeing have used this method.

For the use of its members, IATA publish total loss rates for the six IATA regions, namely:

AFI	-	Africa
CAR/SAM	-	Caribbean / South America
EUR	-	Europe
MID	-	Middle East
NAT/NAM	-	North America / Atlantic
AS/PAC	-	Asia / Pacific

Rates concern only Western built aircraft, principally through lack of available data from former Eastern Bloc states.

Figure 1.1   Regional accident rate per million sectors

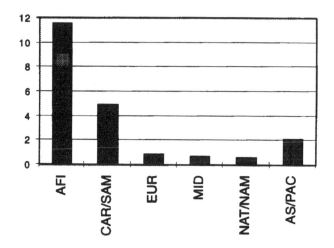

Source: Adapted from IATA, 1993.

The shift in emphasis from million *flights* to million *sectors* takes account of multi-sectoral flights. A flight between two points may involve several sectors for one operator (i.e. several take offs and landings) and yet may be a single sector for another. As approximately 79% of accidents occur in the 18% of flight associated with take-off, initial climb, approach and landing (Boeing, 1997), the use of sectors becomes significant.

The six geographic areas shown in Figure 1.1 cover many countries and therefore many regulators and operators. The question for the researcher is to understand the relevance of the categories used. The regions represent administrative regions for IATA - indeed, until 1994, Egypt and Cyprus were classified as the Middle East, but are now Africa and Europe respectively due to changes in administrative boundaries.

There are obvious differences in accident rates between continents; Africa's accident rate, for example, is nearly twenty times that of North America. However, it would be dangerous to take the inference that all of Africa's airlines have similar safety records any more so than North

American airlines. Analysis of accidents demonstrates that geographical (physical) factors (terrain, weather etc.) account for a very low percentage of accidents. However, the tendency to use such regional statistics is quite common.

Boeing (1993) used 7 regions to describe crew-caused accidents. For 1959-1992, the accident rates are described per million flights.

Figure 1.2   Crew caused accident rate per million flights

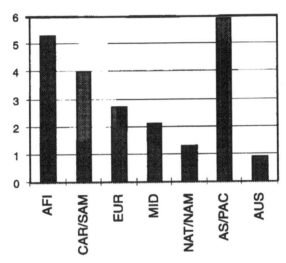

Source: Adapted from Boeing 1993.

Boeing separate Australasia from Asia / Pacific and suggest that this region has the lowest accident rates. Indeed, as Figure 1.2 shows, there is a significant difference between Australasia and the rest of the Asia Pacific region. The crew caused accident rate is over 6 times higher in the latter region than in Australasia. Boeing hints at the role of culture in affecting these human factor accident rates and this effect will be a theme throughout this book. Even at this high level of grouping, the simple changing of groups can affect the apparent accident rates for particular countries. Further disaggregation is desirable if true cultural groups are to be established.

Former American Airlines Senior Director of Safety, Mack Eastburn (1987) presented safety statistics in terms of hull losses as opposed to fatal accidents. His geographical distribution of airlines experiencing hull losses as of January 1987 is shown below.

Table 1.3   Jet aircraft hull loss rate (flying hours) by region

Australia & South Pacific	2,545,000
United States	968,000
Europe	447,000
Canada	582,000
Asia	259,000
Central & South America	201,000
Africa	200,000

Source: Adapted from Eastburn, 1987.

These boundaries are once again different from those used by IATA and Boeing and appear to present Australia in a very favourable light compared to the rest of the world. Eastburn observes that Australia, the best group, is 12.7 times better (in terms of accident rate) than the worst group, Africa. Although he does not offer an explanation of the reasons behind the disparity, he does point out that not only do we know where accidents are occurring, but also what is happening and in what phase of flight. His address was aimed at encouraging safety professionals to use the information available as '...our future success in the prevention of accidents lies in how well we... apply the lessons learned from our past accident experience'.

A note of caution is relevant at this point. Although the accident record shows certain recurrent trends about the types and location of accidents, it is foolish to assume that a good safety record is also to be recurrent. A small number of accidents, especially involving hull loss even with a low number of fatalities, could easily move Australia & South Pacific significantly down the ranking.

There have been a number of attempts to formulate smaller and yet meaningful categories. Oster, Strong and Zorn (1992) created an eighth category in their review of fatal accidents between 1977 and 1989 when Eastern and Western Europe were split. During that period, the authors estimated that Western Europe experienced 1.15 fatal accidents per 1 million departures as opposed to Eastern Europe's rate of 4.11 fatal accidents. As alterations in the membership of regional categories seems to have such a profound effect, there comes a point when countries appear to be the only fair measure of different cultures. This returns to *Flight International's* first statistics and the problems already highlighted.

Table 1.4   Fatal accident rates for scheduled operations by region

Region	No. of fatal accidents	Number of fatalities	% of pax killed in crash	Fatal acc's per 1 m departures	Death risk per 1 m departures
N.America	63	1971	60	0.88	0.53
S. America	64	2326	81	6.04	4.87
W. Europe	30	1663	70	1.15	0.80
E. Europe	33	1662	62	4.11	2.53
Asia	63	2439	64	4.66	2.97
Africa	31	1110	79	13.25	10.52
Middle East	17	1373	69	5.47	3.78
Aus / NZ	5	33	81	1.34	1.09

Source: Adapted from Oster *et al*, 1992.

Oster *et al.*'s approach to regional fatality statistics covers not only the number of fatal accidents per million departures, but also attempts to estimate the risk of being killed on a scheduled passenger flight during the period 1977-1989. (The latter statistic being based on the number of accidents, fatalities and percentage of passengers killed in crashes.) The significance of the death risk seems quite pertinent, especially from the point of view of the passenger. The survivability of a crash may be expected to be a function of several factors including the secondary safety measures afforded by the aircraft (such as crashworthiness, evacuation systems etc.) or airline (emergency procedures) and crew (training for emergencies). If the fatal accident rates can be correlated against aircraft types, operators or regulatory regimes, then the death risk is able to reveal some significant conclusions. However, the authors do not attempt this exercise, possibly due to the small size of the database available.

What is a little confusing though is the fact that the authors claim that the table does not include regional and commuter operations and yet the accidents within Australia and New Zealand appear to reflect aircraft in that class (with an average of 8 seats). The only accident involving a major Australian or New Zealand carrier within the time period 1977-1989 was the loss of an Air New Zealand DC10 in the Antarctic in 1979 with the loss of 257 lives. This is not included in the authors' analyses as it was classified as a charter flight and such flights were excluded. Oster *et al.* also attempted to split the accident record into the fatality rate by carrier for the period 1977-1989. The airlines considered within the region are Air New Zealand, Ansett, Australian, East-West, Qantas, Air NSW and Ansett WA.

Every one of the Australasian carriers listed has had zero accidents and therefore scores a zero death risk per million departures. The accidents that contributed to the previous regional table were obviously not suffered by any of these seven carriers. However, although the authors claim that regional and commuter operations are not included in this analysis, there are no other primary level carriers operating in this area, besides the listed seven. Without examining the raw data used, it is difficult to be completely confident about the analysis.

However, the list of accident and fatality records by carrier provides some interesting statistics to support the view that the Australian carriers enjoy an above average safety record. A selection of carriers are reproduced in Table 1.5. As mentioned previously, the table does not include accidents that occurred on charter flights (even when operated by scheduled carriers). Although the authors acknowledge this caveat, they do not give an explanation of the rationale behind it. This is unfortunate, as the mix of charter to scheduled traffic by different airlines is prone to significant variation. Dan Air, for example, flew predominantly in the charter market, whereas an airline like Qantas or Australian flew predominantly scheduled services.

The effect of excluding charter services is the removal of some very significant accidents including the Pan-Am and KLM B747 collision at Tenerife (1977), the British Airways *Airtours* B737 fire at Manchester (1985) and the Dan Air B727 CFIT accident at Tenerife (1980).

Table 1.5   Safety records for 35 international carriers (1977-1989)

Carrier	Departures (millions)	Fatal accidents	Fatal accidents per 1 mil dep.s	Death risk per 1 mil dep.s
North America				
Delta	7.626	2	0.26	0.13
American	6.035	2	0.33	0.22
USAir	4.774	4	0.84	0.24
TWA	3.515	1	0.28	0.01
Northwest	3.243	2	0.62	0.45
Continental	3.174	3	0.95	0.27
Pan-Am	1.646	4	2.43	1.53
Australasia				
Air New Zealand	0.914	0	0.00	0.00
Ansett	0.903	0	0.00	0.00
Australian	0.414	0	0.00	0.00
East-West	0.300	0	0.00	0.00
Qantas	0.248	0	0.00	0.00
Air NSW	0.190	0	0.00	0.00
Ansett WA	0.183	0	0.00	0.00
European				
Lufthansa	2.778	1	0.36	0.02
British Airways	2.674	0	0.00	0.00
KLM	0.860	0	0.00	0.00
Olympic	0.844	1	1.19	1.19
British Midland	0.448	1	2.23	0.83
Dan Air	0.382	0	0.00	0.00
Asia				
ANA	2.204	0	0.00	0.00
MAS	1.194	1	0.84	0.84
JAL	1.029	3	2.92	1.50
PIA	0.611	3	4.91	3.67
SIA	0.410	0	0.00	0.00
Thai	0.309	3	9.71	9.00
Africa				
SAA	0.732	1	1.37	1.37
Ethiopian	0.179	2	11.17	10.18
TAAG (Angola)	0.005	1	217.39	217.39
Middle East				
Saudia	1.134	2	1.76	0.90
Gulf Air	0.176	1	5.67	5.67
El Al	0.141	1	7.11	0.04
Yemen Airways	0.065	1	15.41	15.80

Source: Adapted from Oster, Strong and Zorn, 1992.

Notwithstanding this, Oster *et al's* study provides a reasonable indication of relative safety performance for many of the airlines listed. Ideally, the greater the number of departures made, the more significant the estimate of death risk can be. However even this ideal is affected somewhat by the time-series nature of the data and safety improvements that may occur within the period of study. Outliers in the data occur when death risk is calculated from the safety record of an airline that has completed a low number of departures and suffered an accident. For example, the single fatal accident, which occurred to a TAAG (Angola) flight within their 5,000 flights, puts the death risk of flying at 217.39 per million departures. While such a fact may be statistically sound for the period 1977-1989, caution must be applied to using it in the predictive sense of measuring safety health.

Even though accident data may suggest that national culture is significant in crafting a safety record, a deeper investigation of different airline safety records indicates that other factors besides nationality are also highly significant. Barnett, Abraham and Schimmel (1979) considered the fatal accident records of 58 major world airlines including 18 US domestic carriers and 40 major flag carriers. Their authoritative text attempted to tackle some of the hazards that had affected previous attempts at classifying safety records. Five of their major concerns were that;

- Different airlines fly routes of different lengths
- Different airlines fly into different airports through different airspace
- Different airlines use different aircraft
- Some airlines have been flying longer than others
- Some of an airline's accidents are probably not its own fault.

In an attempt to rate the airlines' safety records over a particular time frame the authors developed a measure of total involvement in disasters to be measured against the number of flights completed. Although a commonly used measure of accidents is fatalities, such a measure does not take account of the difference between one crash in which 150 die as oppose to three crashes where 50 die in each. Another measure is the proportion of an airline's flights that suffered fatal accidents. However, this ignores factors such as passenger load and how many survived as a result of airline-controlled factors such as the quality of emergency procedures.

The solution was to produce an alternative safety measure called the "cumulative fatality quotient" (CFQ). The fatality quotient (FQ) is defined as the fraction of passengers who do not survive any given flight and the CFQ incorporates the total number of flights studied. From this figure, the authors derive the "average fatality quotient" (AFQ) as a measure of over-all safety performance, or in other words, the probability that a person would die on any flight during the studied period chosen at random.

The 18 US domestic carriers were considered first and these results are considered to be outside the remit of this book. However, the authors provide a good demonstration of the dangers of regionalising statistics. By converting the calculated AFQ values, the odds of being killed on a US domestic flight are shown in Table 1.6.

Table 1.6   Death risk for US domestic flights

Period	Odds of being killed on a flight
1957-61	1 in 988,000
1962-66	1 in 1,087,000
1967-71	1 in 2,064,000
1972-76	1 in 2,599,000

Source: Adapted from Barnett, Abraham and Schimmel, 1979.

These results clearly demonstrate the considerable improvement in aviation safety over the 20-year period and yet also show a clear disparity between the US and the rest of the world (Table 1.7).

Table 1.7   Death risk for rest of world flights

Period	Odds of being killed on a flight
1960-64	1 in 163,000
1965-70	1 in 366,000
1971-75	1 in 340,000

Source: Adapted from Barnett, Abraham and Schimmel, 1979.

Acknowledging the constraints stated at the outset, Barnett *et al.* produced a table of accident records as shown in Table 1.8.

Table 1.8   **Safety records for 40 international carriers (1960-1975)**

Airline	No. of flights (000's)	Expected CFQ	Actual CFQ	Death risk per flight
Aer Lingus (Ireland)	430	1.62	1	2
Aeromexico	74	0.23	1.05	14
Air Canada	600	2.23	0	0
Air France	1379	5.01	6	4
Air India	132	0.47	1	8
Alitalia	883	3.04	1.86	2
Aeronlinas Argentinas	165	0.59	2	12
Austrian	207	0.74	0.84	5
Avianca (Columbia)	77	0.26	0.79	10
British Airways	2180	7.95	5.04	2
East African	103	0.36	0.34	3
Egyptair	156	0.57	6.02	39
El Al (Israel)	117	0.41	0.02	0.2
Ethiopian	102	0.39	0	0
Finnair	228	0.82	0	0
Iberia (Spain)	549	1.67	2	4
Iranair	68	0.21	0	0
Japan Airlines	315	1.00	1.79	6
JAT (Yugoslavia)	187	0.61	1	5
KLM (Holland)	1010	3.82	1.55	2
LAN (Chile)	67	0.24	1	15
Loftleider (Iceland)	52	0.19	1	19
LOT (Poland)	124	0.45	1	8
Lufthansa (Germany)	1032	3.50	0.38	0.4
Malev (Hungary)	44	0.17	3.88	88
Nigeria	49	0.18	1	20
Olympic (Greece)	161	0.58	0	0
Pakistan International	97	0.33	1	10
Pan American	1897	7.68	5.61	3
Phillipine	41	0.15	0	0
Qantas	247	0.88	0	0
Sabena (Belgium)	611	2.36	1.07	2
SAS (Scandinavia)	1177	4.35	1.33	1
South African	110	0.38	0.96	9
Swissair	995	3.56	2	2
TAP (Portugal)	116	0.39	0	0
THY (Turkey)	81	0.28	1	12
TWA	385	1.34	1.60	4
Varig (Brazil)	139	0.46	2.57	18
Viasa (Venezuela)	85	0.34	2	24

Source: Adapted from Barnett, Abraham, Schimmel 1979.

Death risk can be seen to vary between zero and 88 (per million flights) with Qantas being rated as zero. Closer examination by the authors revealed an apparent size dichotomy where larger airlines experienced significantly fewer than predicted accidents and smaller airlines, the reverse. However, even though Qantas and El Al (Israel) are classified as small airlines, they stand out as exceptions to the rule. Barnett *et al.* speculate that this may be that the key safety factor is not the size of an airline but the level of '...technological development of the mother country'.

In 1993, the International Airline Passengers Association (IAPA, 1993) completed a safety rating exercise using ten years of data in an attempt to '...predict the future risk of a particular airline having a fatal accident'. Six months of research led to the formulation of a method to rate the relative safety of the world's airlines. This included both the more easily quantifiable factors such as accidents per million flights or deaths per million passengers and more difficult factors such as the quality of pilot training and maintenance. Unusually, it also considered other factors such as value of fleet, age of aircraft, its country's air traffic infrastructure and government commitment to safety. Although IAPA conceded that they rated certain airlines as safe, they could not guarantee that they would not have a fatal accident in the future. There was also limited explanation of the methodology or the criteria used to assess the qualitative factors.

Airlines were classified by size as shown in Table 1.9. A month after the original list was published in August 1993, IAPA added to the list the four small airlines it considered to be the safest. In alphabetical order, these were Cathay Pacific, Qantas, Sabena and Singapore Airlines. All four '...are in politically stable countries, have a long history of excellent management and take their airline's safety very seriously.' (IAPA, 1993b)

Although Ansett and Qantas are mentioned as being among the safest, such a statement should only be seen as corroborating the work of Barnett *et al.* and not a stand alone judgment. The lack of explanation regarding either the selection of classifications or the quantification of qualitative data does little to add to the credibility of the statistics. There were a number of apparent oversights that deserved further investigation.

For example, the statement concerning SAS being accident free for over 24 years seemed to ignore the occasion in 1991 when an MD-80 aircraft suffered a double engine failure shortly after take-off and crashed in a nearby field. Although the aircraft broke into three parts and was a total hull loss, there were no fatalities. IAPA argued that its definition of an accident

Table 1.9   IAPA honour roll of safest airlines (1993)

Group A: The Best of the Biggest

Qualifications - Over the past ten years:

1. *At least 2,000,000 flights*
2. *Less than one accident per 2,500,000 flights*
3. *Less than one fatality per 4,000,000 passengers*
4. *Fleet age under 12 years*

American Airlines	over 9,000,000 sectors without loss
British Airways	one accident in 3,000,000 flights
Delta	one accident per 4,000,000 flights
Lufthansa	one fatality per 20,000,000 passengers
Scandinavian	accident free for over 24 years
Southwest	largest airline in the world never to have an accident

Group B: The Best of the Mid-Sized Airlines

Qualifications - Over the past ten years:

1. *1,000,000 to 2,000,000 flights.*
2. *Zero accidents.*
3. *Fleet age under 12 years.*

All Nippon Airways	no accidents in 22 years
America West	no fatalities
Ansett Australia	no jet crashes in over 25 years
Canadian Airlines	only 1 accident with more than one fatality in last 25 years
Saudia	12 years since last accident

Group C: The Best of the Medium-Small Airlines

Qualifications - Over the past ten years:

1. *600,000 to 1,000,000 flights*
2. *Zero accidents*
3. *Fleet age under 12 years*

Alaska Airlines	one fatality 17 years ago
Finnair	no jet crashes ever
KLM	no accidents in 16 years
Malaysian	no fatalities in 15 years
Swissair	

Source: Adapted from IAPA, 1993.

required there to be a fatality (contrary to the ICAO and IATA definitions) because '...IAPA feels that what passengers care about most is fatalities.' (Burchill, 1994) Although IAPA eventually conceded that the accident was

'...100% avoidable and a disgrace to modern aviation' (Margolis, 1994), the fact that such an accident had no different a set of causal factors than a fatal accident casts a shadow on the reliability of such statistics for assessing the true safety health of an airline.

Ironically, a number of the airlines listed in the IAPA study went on to have fatal accidents including American, Saudia, Alaska Airlines, Singapore Airlines and Swissair. Such is the hazard of measuring safety health based on fatal accidents.

Notwithstanding this, all of these statistics demonstrate a similar theme, which is that although it would be unwise to take the rating of airlines as being a guarantee as to their future safety, they do give some indication of safety performance. What is also true is that the principal Australian carriers consistently appear as *above average*. It would be quite pointless and rather foolish to try and prove that the Australian carriers are the safest in the world. Movie watchers may have delighted in the spectacle of the autistic Raymond from the film *Rain Man* refusing to fly on any airline other than Qantas as 'Qantas never crash...' but such an assertion is not open to scientific proof and neither should it be.

The fact is, as *Flight International's* Learmount put it in 1987, that; '...anybody in the air transport business will recognise that Qantas is operationally excellent', but as soon as anyone tries to prove that they, or indeed any other operator, are *the safest* then they enter the realms of pure speculation. Statistics allow researchers to see that the safety record is above average and offer lessons for operators all around the world.

This analysis confirms the difficulties of using statistics to examine both safety records and the safety health of any airline or country. Nevertheless, they also seem to agree that the record for Australian carriers is above average by virtue of zero hull losses in commercial jet RPT aircraft. Detractors of the record would point out that should Australia lose a large aircraft such as a B747, then the statistics for Australia's safety record would change significantly due to the comparatively low number of sectors flown and the current zero loss score. While this may be true in the pure statistical sense, such logic would seem to suggest that the only way to measure the safety record accurately was if there was a crash to divide the number of sectors flown by. In other words, zero hull losses is not a good number for statisticians to work with!

Even if such an accident occurred tomorrow then the important question to answer is, would such an incident negate the value of this research? The answer is most likely no as this work can only hope to cover what has created a good safety record so far. It can also advocate strengths that can be maintained and weaknesses that must be addressed, but it cannot guarantee a continued good safety record.

Flight International (Taylor, 1988), Boeing (1993), Eastburn (1987), Oster, Strong and Zorn (1992), Barnett, Abraham and Schimmel (1979) and IAPA (1993) all produce evidence which puts Australian aviation or its main airlines (Qantas, Australian and Ansett) at the top of the list in terms of safety record. Even though some of Oster, Strong and Zorn's (1992) data seems to suggest the Australasian rate is worse than North America, they do not present sufficient data to explain this relative to their list of seven carriers from the region, all whom have a fatal accident rate of 0.00 per one million departures. ICAO chooses not to release comparative safety statistics and the IATA statistics are not useful for this case study by virtue of Australia being included into the diverse *Asia Pacific* region.

It is the author's contention that none of the statistical techniques that are currently available are sufficient to be used exclusively to answer the question "Is Australia's safety record good?" However, as they all seem to say similar things about the Australian record, it seems fair to assume for the purposes of this book that Australia's record for jet RPT operations is above average or *good*. This supports the use of Australia as a case study to examine the reasons behind safe aircraft operations.

2 Defining Safety

How do we Define Safety?

Safety is an important issue for aviation. For some, this is because aviation has the potential to be so dangerous, whilst for other it is because aviation has proven itself to be so safe. Either way, *safety* is condition to be striven for, but does anyone really know what *safety* actually is? In the absence of a universal description, it is possible that everyone is actually aiming towards different goals.

The Collins English Dictionary defines safe as being 'the quality or state of being free from danger', but such a loose description allows many interpretations. If *safe* is used in the absolute of form of meaning free from any danger then it can only really be directed towards a simple cause and effect relationship. For example, if a baby and dog are kept together in the same room then there is a danger that the baby could be attacked by the dog. By removing the animal from the room, the baby may be considered to be free from that danger and therefore safe. However, this does not account for the fact that there are other dangers in the room such as falling furniture or electrical appliances. In the wider sense, the removal of a particular risk is rarely a guarantee of absolute safety. Real life situations will be prone to a variety of risks - the absolute definition of safety would suggest that if this is true then it is all but impossible to achieve safety.

Safety is not measurable - risks are. Safety may be judged relative to its level of risk versus the acceptable level of risk. To determine safety therefore, involves two quite separate activities; measuring risk and judging safety i.e. the acceptability of risks. It is therefore vital to reconcile the term safety with risk.

A deeper investigation of the meaning of safety is not a new thing. Slovic, Fischhoff and Lichtenstein (1980) remarked on the '...widespread confusion about the nature of safety' adding that '..for a concept so deeply rooted in both technical and popular usage, safety has remained dismayingly ill-defined'. Their chosen definition of safety - 'a thing is safe if its risks are judged to be acceptable' acknowledges the impossibility of eliminating all risk and adds the relative and judgmental nature of safety. Safety is a state that may be defined by individuals and is therefore potentially variable in its definition. Some definitions of safety refer solely to human life. Jones-Lee's (1989) exploration of safety and decision making refers exclusively to 'the safety of human life and the degree of protection from physical risk'. The Royal Society (1992) clearly states its understanding of the concept of safety as:

'Safety relates to the freedom from risks that are harmful to a person or a group of persons, either local to the hazard, nationally or even worldwide. It is implied that for the consequences of an event to be defined as a hazard, i.e. a potential for causing harm, there is some risk to the human population and therefore safety could not be guaranteed, even if the risk is accepted when judged against some criterion of acceptability'.

Safety is a delicate balance - an ideal state and yet one where the true condition can only be known in retrospect. Experience and experimentation lead to a level that may be labelled as *safe* and yet this is a label that can only really be used after the event. This is a function of two things - the problem of defining an *acceptable* level to be labelled *safe* and the variability or unpredictability of certain risks. Before it is possible to understand these problems, it is necessary to explore the concept of risk in more detail.

Risk is a word of many meanings: Although it is relatively simply defined by the dictionary as 'chance of bad consequences or loss', it is a word that is used in different contexts to mean different things. It is therefore important to develop a working definition of risk for use in the context of this text.

Solomon (1993) points to three quite different contexts in which the concept of risk occurs;

1. In a decision-analytical sense, risk refers to the fact that the consequences of choosing an alternative are not known with certainty, but instead can be expressed as a probabilistic outcomes. In this sense, no reference needs to be made about the positive or negative effect of the consequence.
2. The popular view of risk emphasises the probability of a potential harm and focuses on the probability of that harm without regard to the (negative) magnitude. The benefits in this sense are completely ignored.
3. Between these two viewpoints lies a third definition of risk - the probability of an event times the consequence of an event summed over all events.

The Royal Society's (1992) definition is closest to the third of Solomon's statements; 'risk is a combination of the probability, or frequency, of occurrence of a defined hazard and the magnitude of the consequences of the occurrence'. Both definitions clearly indicate the components of magnitude and probability as being critical to any risk.

Previous to the Royal Society, the Council for Science and Society (CSS 1977) also explained the concept of risk relative to hazards. A hazard describes a situation with 'the potential to cause harm (to people, property or the environment)' and risk in turn refers to 'the probability that the potential hazard will be realised'. This definition accepts that the consequence of a hazard (and realised risks) can be more than the human loss that is so often referred to.

All of these definitions suggest that some measure of risk is possible in the form of an objective quantitative probability. In its absolute sense then this may be true, but the wide use and misuse of the term *risk* raises the possibility of it being used to mean rather different things.

Risk may be considered as the possibility of misfortune or loss and is based on probability (whether calculated or perceived). 'All activities have an inherent risk and risks can never be removed completely from any activity.' (Solomon, 1993) In common usage, the word "risk" is generally associated with "high risk". Phrases that describe an activity as being "open to some risk" or an individual's actions as being "risky" suggest that the risk is higher than normal. All activities, by definition, carry some element of risk. The confusion appears to lie in discriminating between risk and safety. Activities that are deemed to be safe still carry an element of risk. The presence of risk does not necessarily equate with a lack of safety.

There is no unique measure of risk because there are so many different types of exposure to risk. For example, whereas all the population are exposed to cosmic radiation and therefore, the risk of skin cancer can be calculated per group of population, only a limited number of people are exposed to, say, the risks of a motorcycle accident by virtue of the fact that only a small population ride motorcycles.

The measure of accidents can also be quite different. While fatalities may seem to be a reasonable and easily defined measure, it can also give a false impression. For example, as Solomon points out, one hundred car accidents each with a single fatality is not perceived to be equivalent to a single accident that kills one hundred people. Even classification of accidents by fatal, serious and minor injuries may be misleading as it ignores the quality of life issue. As human life can not be valued (except by arbitrary figures for cost benefit analysis), it is even more difficult to try and suggest a value for a serious injury than for a fatality. For example, the fatality rate for accidents in paper mills may be low compared to that of air travel but there may be a significantly higher number of maimings and serious injuries in the former. Therefore, fatalities as a stand-alone measure may prove to be misleading as a measure of overall risk.

The risk measure must therefore be sensitive to the nature of the particular risk it is designed to represent. If the risk relates to catastrophic events such as dam bursts or ferries lost at sea, it is sensible to measure the probability that more than a given number of people are killed in an accident in a particular time period. According to Solomon, this provides a useful measure of distribution for comparing the way they are perceived psychologically. An alternate measure is the expected number of fatalities within in a particular group over a specified time. For example, airline passenger fatalities per 100,000 hours flown.

Slovic, Fischhoff and Lichtenstein (1980) suggest evidence regarding hazards may be gathered from a number of sources such as:

- Traditional or folk knowledge
- Common-sense assessment
- Analogy to well known cases
- Experiments on human subjects
- Review of inadvertent and occupational exposure
- Epidemiological surveys
- Experiments on non-human organisms
- Tests of product performance.

Johnson and Covello (1987) discussed the social construction of risk highlighting the issues behind '...the claim that risk is a social construct stemming primarily or wholly from social and cultural factors'. One perspective is that this implies '...that people are incapable of perceiving what is really dangerous since there are no actual or objective risks in the world.' Risk perception research sees the convergence (and even divergence) of psychology, sociology and politics in an attempt to explain a process that is governed by '...knowledge, values and feelings and is considerably dependent on the cultural / societal context.' (Rohrmann, 1995)

Rohrmann observed that the concept of risk has become more topical during the last decade. In particular, the hazards perceived to be posed by large-scale technologies such as nuclear power and chemical industries have led to an increased awareness of accidents (acute threat) and longer term environmental or health damage (chronic threat). Add to this an increased awareness of the human sciences of psychology, physiology and sociology and the result is an increasingly complex societal decision making process.

It is the subjective nature of risk that demonstrates the role of society in creating and enhancing the concept. Regardless of the efforts of international human rights activists, the fact remains that the price of human life still varies significantly around the world. This means that identical risks may have different values depending on the 'value' of the lives involved.

'The term risk acceptability conveys the impression that society purposely accepts risks as the reasonable price for some beneficial technology or activity.' (Kasperson and Kasperson, 1984) Logically this is exactly right; society does accept risk and yet, as a result of confused usage of the term, there is a perception that risk is not acceptable in any form. Kasperson and Kasperson continue to suggest that in respect to the above statement; '...For some special cases this may approach reality. Hang-gliding, race-car driving, mountain climbing, and even adultery, divorce, and midlife career changes are all high-risk activities in which the benefits are intrinsically entwined with the risks. These activities are exhilarating because they are dangerous. But most risks of concern are the undesired and oft unforeseen by-products of otherwise beneficial activities or technologies.'

The Norwegian safety society, Det Norske Veritas, provides a clear explanation of the process of risk acceptability, 'The criteria for the acceptance of risk in relation to the safety of human life spring from each person's attitude towards safety and also from social conditions and the attitudes of the mass media and politicians. We are here faced with a spectrum of attitudes. Some of these may serve as starting points for rational evaluations and decisions: other attitudes are too narrow or emotional to provide general norms.' (Det Norske Veritas, 1979)

'The acceptable level of risk is not the ideal risk. Ideally, the risks should be zero. The acceptable level is a level that is good enough, where good enough means you think that the advantages of increased safety are not worth the costs of reducing risk by restricting or otherwise altering the activity.' (Slovic, Fischoff and Lichtenstein, 1980)

The notion of "affordable safety" seems to bring shivers to the collective spines of professional aviators throughout Australia. This is largely a function of the man who first popularised the phrase and his interpretation of it. During his time as Chairman of the Civil Aviation Authority Board, Dick Smith was responsible for a wide-scale restructuring and down-sizing of the authority which was very unpopular with a large cross-section of the aviation industry. 'Those who object to the term affordable safety saw it as a key feature of the period they considered to be the dark ages of aviation safety regulation in Australia or the slash and burn period of safety regulation. This was the period when Smith was Chairman of the CAA.' (HORSCOTCI, 1995)

The House of Representatives Standing Committee explains affordable safety as being a concept with two separate parts:

1. Changes and standards are market driven and if the costs of regulatory changes are too high, excessive ticket prices will result in fewer people flying and more people using less safe road transport.

2. There is a finite safety dollar which, in the case of the CAA meant reallocating resources to maximise lives saved.

The House of Representatives Standing Committee concluded that there was nothing objectionable about affordable safety although they conceded that this was not a term they would use. In fact, the whole concept of safety is based upon affordability, that is the point at which we consider certain risks to be acceptable. i.e. where we consider resources allocated to risk countermeasures to have reached the reasonable limit. Nevertheless, the concept of affordable safety as defined above is worth deeper consideration.

The concept of the market regulating aviation safety is based on economic forces. These balance the cost of risk countermeasures against the risk and cost of accidental consequences. However, Smith's argument seems based on the actions of the end user i.e. the passenger - the very section of the community who are the least informed about aviation safety matters. Chalk (1987) suggests that '...although consumers lack the technical expertise to measure the inputs to safety provision, they can measure the output - fewer accidents. If consumers are informed about accidents for which an airline or an aircraft manufacturer is considered culpable, they can avoid riskier airlines and aircraft and manufacturers.' The crucial question is who is responsible for the communication of critical safety related information and how is its integrity assured?

The mass media is the principal conduit for information and brings with it the great variations in reliability that are associated with a communication method whose success is measured in sales or viewing figures. An example of the variations would be the stories that have been published regarding the cause behind the loss of TWA 800 off Long Island in June 1996. Stories have circulated and been aired which talk about terrorist devices, a "friendly-fire" missile or an explosion in the centre wing fuel tank. Passengers could easily come to the conclusion that airport security in the USA was deficient, US airspace unsafe, or the B747 a defective aircraft.

Even on the single issue of the aircraft suffering a catastrophic in-flight failure, will the average passenger be provided with enough facts to decide whether the failure was a result of poor maintenance on the part of Trans World Airways, a design fault on the aircraft or a result of the aircraft being 25 years old? Indeed, if they decided that the problem was a design fault on the B747 then how many passengers would know the difference between the Classic (-100, -200 and -300 series) and Advanced (-400) series aircraft? As the B747 is the mainstay of international operations is it reasonable to believe that such a process is affecting modal choice?

The acceptability of risk is not the same as the elimination of it. Everyone accepts risks in one form or another as part of daily life. The crucial factors that set skydiving grandmothers apart from those who sit darning socks and, moreover, those which appear to separate "right stuff" aviators from the average airline pilots, seem to be the level of acceptable risk. What is most important to the integrity of the aviation industry is that the decision makers, at whatever level, are armed with an accurate perception of risk acceptability. Slovic, Fischhoff and Lichtenstein (1980) note that 'people respond to the hazards they perceive' and consequently if these '...perceptions are faulty, efforts at public and environmental protection are likely to be misdirected.'

Safety, based upon the concept of risk acceptability is perhaps best explained in the form of a very simple diagram, as shown in Figure 2.1. Absolute safety is achieved when risk exposure is balanced perfectly with safety measures. This is a Utopian system that can never be achieved as it relies on complete knowledge, not only of the full set of risks which may occur, but also the full properties of safety measures. It has never been, and will never be, achieved in any system that has a human involvement. Such is the inherent nature of human error. The accident record shows events occurring because of risks that were unforeseen or the full effect of actions or procedures were unknown.

Figure 2.1   Absolute safety

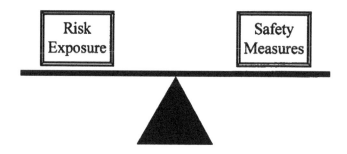

Absolute safety represents a perfect economic system where the amount spent on risk-countermeasures is precisely balanced by the savings on accidental losses. In accepting that this ideal is a point to aim for, safety professionals attempt to create a system as close to this as possible. This process is an attempt to achieve the point of risk acceptability - an area of balance where the wastage is minimised.

A more achieveable goal for safety professionals is that of acceptable risk, where the level of risk exposure is slightly greater than the level of safety measures. If all of the risks were to present themselves then accidents would still occur, but as this is unlikely, acceptable risk represents a situation where the safety measures which are selected represent the best guesses or most effective measures. The safety system will cover the most common events or most dangerous events, but this may still mean that some accidents will occur. The clever science is in predicting the risk exposure profile accurately and preparing a set of countermeasures which adequately caters for it.

If the level of risk exposure is allowed to become too great then the balance will tip towards unsafety. Conversely, if the safety measures are excessive then the balance will tip towards a condition of *over-safety* where economic efficiency is lost. In the greater scheme of things this may mean new safety problems for other systems elsewhere. For example, if aviation became so expensive for this reason, then travellers may switch to modes such as car travel with inferior safety records, which seem cheaper.

It is, however worth considering that there is a point on the safety balance where acceptable safety and *over-safety* start to meet and where resource allocation affects the margin of safety that is often afforded in complex systems. The extra safety that is often included in a system *for a rainy day* or to *round up* safety calculations may be responsible for catching some of the outlier accidents, which would have occurred if the balance of acceptable safety had been more strictly enforced. It is possible that part of Australia's good safety record is as a result of this additional safety margin - decisions that may have appeared uneconomic at the time but in retrospect have been useful and necessary measures.

What is clear is that there are many interpretations of what is safe. If it is assumed that safety is the condition where risk is controlled to an acceptable level, then the problem of defining the term safety is shifted to that of defining the level of risk that is acceptable. This concept will be explored in more detail throughout this text, but it is important that the reader understands the difference between safety and risk at an early stage.

It is suggested that Australian aviation may possess a different perception of acceptable risk than other countries / authorities / airlines around the world and it is important that these differences are examined, if they do in fact exist. In simple terms, the following issues will be considered;

- Do Australians possess a lower acceptance of risk?
- Does Australian aviation possess a lower acceptance of risk?
- Do Australia's airlines possess a lower acceptance of risk?
- Does aviation recruit staff who possess a lower acceptance of risk?
- Does Australia have the same risk acceptability as other countries, but a lower exposure?

Measuring Safety

It has already been suggested that safety cannot be measured, as safety is an outcome and not a state. It is therefore important to make a distinction between attempting to measure safety (as a state) and measuring safety records. A number of attempts have been made to do the latter (which is valid when used to look at a historical outcome) and most attempts to do the former have ended up being a disguised version of the latter! Measuring safety records can give trend information, but can also be misleading.

The production of safety statistics is a prolific industry by itself, based upon the premise that a record of accidents or incidences is the measure of how *safe* a particular activity was. The most commonly used measures are in terms of fatalities or simply events over time.

Fatalities, (or often fatalities per 100,000 persons per year) is a favoured measure because of the requirement to report such occurrences, usually through the Coroner's office. This allows reliable statistics to be collated for analysis on an international level. Deaths are classified by an International Standards Organisation code which cover all but the most bizarre causes of death. Such records exist for a number of years and often in computerised form to demonstrate data trends and smooth out statistical noise.

Mortality rates for the general public that include death from natural causes (disease and natural disasters) put the risk from accidents into perspective. Solomon (1993) documents statistics on mortality from the Universal Almanacs as follows;

Table 2.1   **Typical mortality rate (per 100,000 persons per year) for the US general public, 1965-1992**

Death from disease	860
Death from all accidents	54
Death from natural events	0.84
Total risk	**930**

Source: Adapted from Solomon, 1993.

Heart disease accounts for well over a third of all deaths (360 per 100,000 persons per annum) or well over six times as many as from all accidents. However, safety statistics are rarely grouped in such a way as to bring the full spectrum of hazards together. Accidental deaths are grouped together and then often split into occupational hazards, transport, sport and so on. Some would argue that accidents should be separated because they are, by definition, preventable. However, many of the deaths due to disease are also preventable (e.g. some heart and lung diseases due to smoking) as are some of the deaths from natural events. (e.g. those living on flood plains)

The problem with comparing safety records between different modes of transport is finding a fair measurement to use. Some modes are used more than others, cover greater distances and operate at different speeds. Further, the level and severity of injury experienced from certain transport modes is highly variable e.g. slow speed modes like bicycles attract more injury accidents than fatalities in contrast to, say, aircraft travel. To compare air with bicycle travel over a fixed mileage would favour air travel unless modes were compared solely on the basis of fatal injuries.

Solomon observes the level of accidental deaths between transport modes in the USA to conclude that flying is very much safer than car travel in terms of accident rate per one hundred million passenger miles. The results of his statistical analysis are presented in Figure 2.2.

Figure 2.2   Accidental transportation deaths per 100 million passenger miles

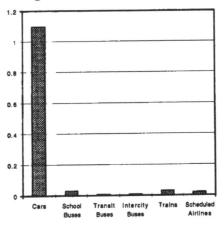

Source: Redrawn from Solomon, 1993.

Reason (1993) however, provides a compelling warning to those accepting this apparently safe situation at face value. If the average failure accident rate for commercial aviation is taken to be approximately one per million flying hours, this is a mortality rate 2-3 times worse than the UK accident risk in cars since wearing seat belts became mandatory. In turn, aviation has a mortality rate sixty times that of passenger trains. Car journeys are generally short in nature (compared to air travel) and accidents are often non-fatal. Railway accidents are relatively infrequent and have a higher degree of survivability than air accidents.

Woodhouse and Woodhouse (1997) argue that the most reliable indicator of risk in aviation is accident rate (accidents per million sectors). They suggest that, 'Accidents are a more appropriate numerator than either fatal accidents or fatalities because the survivability of accidents is so much a matter of chance.' However, they also add a cautionary note regarding the existing databases of aircraft accident statistics. 'Until 1993, notification of accidents to ICAO was a recommended practice rather than a standard. ICAO data is therefore incomplete, particularly for serious damage accidents'.

Waldock (1992) wryly observes the problems of observing safety in aviation even through the collection of accident statistics; '...they are rare events; they don't happen frequently enough to generate a large enough population to be valid. One bad accident can skew the statistics drastically. If we had more accidents to work with, we could probably do a better job of measuring safety by statistical analysis'.

Reporting of Occurrences

The integrity of statistics is fundamentally controlled by the quality of the data from which they are sourced. In other words, safety statistics are only of any real use if they represent the full picture. This is one of the reasons why transportation safety statistics tend to use only fatalities to compare records between modes. While fatal accidents must, by law, be reported to the coroner in most countries; serious injury or minor injury accidents do not need reporting. Therefore statistics which claim to describe all accidents may be incomplete because of this reason.

Although ICAO *Annex 13* requests the reporting of all aircraft accidents, it is apparent that certain member states have not followed this recommendation. Accidents within former Soviet states and China are known to have occurred and not reported. For example, in one accident a B737 was lost and the airframe written off by the insurance company although is yet to be reported as an accident to ICAO. For this reason, aviation safety statistics tend to exclude losses from these areas and to Eastern-built aircraft types.

Incident reporting is a safety measurement tool, which is being increasingly used around the world. However, it remains a source of confusion as a result of the way such reporting systems are set up. Incidents may mean a very near accident - this may be a spectacular event with lots of witnesses such as a near-miss in mid-air or a go-around executed to avoid a runway incursion, or they may mean an invisible event such as flight crew falling asleep in flight or a single pilot nearly colliding with terrain outside controlled airspace. Although some countries such as Australia do have a mandatory incident reporting system, many do not and therefore depend on voluntary reports by flight crew or air traffic controllers. This is why databases tend to be incomplete. Whether the reporting system is mandatory or voluntary, the crew or controllers involved need to recognise that they have had an incident and then, in the case of voluntary systems, need to believe it important enough for them to fill in the required paperwork. This is often at the end of a long trip when crewmembers would rather go home if they can get away with it. Individuals may also wish to consider their legal position in filing a report.

Although airline and national systems such as Australia's CAIR system or NASA's ASRS systems are run as confidential and non-punitive systems, there is growing concern that any safety critical information which is recorded may later be subpoenaed in a court of law. This is because although agreements often exist between the aviation regulators, airlines, unions and crews; Federal courts can still sometimes overrule them. This issue recently presented itself in New Zealand following a controlled flight into terrain (CFIT) accident where the flight crew were charged with manslaughter based on a cockpit voice recorder (CVR) transcript. Even though the judge eventually ruled the recording inadmissible as evidence, the threat of similar events has been enough to cause incomplete reporting of incidents. Further to these reasons, there is also the simple fact that individuals may choose not to admit to an incident because they are embarrassed or ashamed and don't want anyone else to know.

As the ways to report incidents improve such as through better access to, or better designed reporting forms, so too the number of incident reports will rise. This has caused some consternation in organisations as incidents appear to escalate. The crucial distinction is that it is generally only the number of reported incidents that have increased, rather than the absolute number of incidents. This is actually a good thing in terms of organisational safety. British Airways is proud to report an average of 2,000 reports a year for its BASIS confidential reporting system and contrast this against a certain US major airline which claims to have no incidents to report.

Unless complete incident reporting can be guaranteed (which is unlikely), the number of incidents is only a very rough safety measurement tool. Even if the absolute number of incidents is ignored for the reasons stated above, and analysis is restricted to the proportion of incident types, this is still prone to a level of error. It assumes that those reporting the incidents are able to accurately spot and assess the severity of incident types. It may be that the most dangerous proportion of incidents are those that remain unrecognised and therefore invisible in published incident statistics.

The Customer and Measures of Safety

Although accident statistics are collected and prepared by a number of agencies, it is important to recognise the potential for these figures to be misquoted or used selectively outside of aviation. The customer's measures of safety do not follow rigid rules of academic rigour and can be influenced by a number of subtle processes. One of the strongest influences comes from the media, which is something that must be borne in mind when tackling the subject of safety measurement.

Safety is generally negatively reported. In other words, it is only mentioned in times of un-safety. This is especially so in the mass media where accidents sell newspapers and increase television ratings. Coverage of accidents appears to be proportionate to the visibility of the event. Large railway accidents in which only a single person dies often command far more coverage than small car accidents where two or three die. For example, in 1985, two passenger trains collided at Colwich Junction near Stafford, UK with the loss of one of the drivers. The crash scene was spectacular and the story covered most of the first six pages of the tabloid newspaper *Liverpool Echo* (Liverpool being the destination of one of the trains).

On the seventh page, a story about the death of two locals in a car crash filled three inches of single column space.

Aircraft accidents in particular seem to attract a disproportionate amount of attention because they often involve a high number of fatalities in a short period of time and leave a large amount of wreckage. The presence of explosions and fire only heighten the visibility or reportability of such events. The relative novelty of mass travel by air means that many generations still retain a mystified view of aviation, which both propagates and in turn is propagated by the mass media.

The media's prime task is to bring fast and up to date information to its customers which is often contrary to the objectives of safety investigators who are trying to ensure integrity of information. In the first hours and days following an aircraft accident, there is often mass speculation, which can mislead the public. By the time accident investigators have done their job and produced a complete accident investigation report, the public interest has often waned. Hence, misconceptions formed at the speculatory phase tend to live on in the public's perceptions.

The crucial point is that although the media is capable of leaving the public with a misdirected perception of safety matters, it is that perception that ultimately influences their decision making. Curtis (1996) observed the way aircraft accidents were reported in the New York Times between 1978 and 1994 to examine the nature of media bias. The conclusions were that jet aircraft events were more likely to be reported than corresponding propeller aircraft events (categorised by fatalities) and that although fatal accidents were reported with a greater magnitude, there was no clear relationship between the amount of coverage and the number of fatalities. More specifically, Curtis suggests that the reporting of deaths caused by hijack, sabotage or military action is disproportionate high enough to leave readers with a distorted view of the hazard.

The influence of the mass media can also work in reverse albeit with less regularity in practice than a negative effect. The 1987 film *Rain Man*, starring Tom Cruise and Dustin Hoffman, left the public with a "rose-tinted" view of the integrity of Qantas over other international airlines. This view continues today, nearly 15 years after the film's release.

During the course of this research, non-aviation experts have made much comment. Most people have heard of Qantas as being the safest airline in the world and then quote the source of knowledge as the film. If Hollywood says that 'Qantas never crash', then it seems, that may be good enough for the man in the street.

However, it is also worth noting that the influence of the media can also be positive. Disproportionately high levels of attention on aviation safety heighten consciousness and awareness. As such this is not always a bad thing for future progress. Funding for safety measures, research or additional safety staff is often much more forthcoming following a serious accident. The proportion of research money spent on evacuation trials at Cranfield in the UK in the aftermath of the British Airtours accident at Manchester in 1985 is one example. As one senior airline manager commented during this research, '...you should see the safety budget grow in any airline following an accident'.

The effect of biased reporting of aviation accidents and incidents may be one of the reasons that it is such a safe mode of transport. The inaccurate communication of risk may be one of the reasons that funding has been available to elevate aviation safety to such a prominent position.

3 A New Approach to Safety?

Why Do Accidents Happen?

The Collins English Dictionary (1993) defines an accident in three different ways as;

- an unforeseen event or one without an apparent cause;
- anything that occurs unintentionally or by chance;
- a misfortune or mishap, especially one causing injury or death.

Such definitions need some alteration before they can be applied to the aviation environment. For example, there are few accidents which cannot be assigned *an apparent cause*. One such definition would be that used by Boeing (1996) in the preparation of its annual accident summary

> Aircraft accident means an occurrence associated with the operation of an aircraft that takes place between the time any person boards the aircraft with the intention of flight until such time as all persons have disembarked, in which any person suffers death or serious injury as a result of being in or upon the aircraft or by direct contact with the aircraft or anything attached thereto*, or the aircraft receives substantial damage.

> * Includes the effects of jet blast but not of wake turbulence.

Accidents as System Failures

In the context of this work, an accident will be considered to be an unforeseen system failure and an incident, a partial system failure. The aviation system is far more than just a physical set up, and is best classified as a *sociotechnical system*. A pipework failure study conducted by Technica (1992) demonstrated a sociotechnical system to be comprised of five levels.

The pyramid of components shows increasingly remote causes of accidents demonstrating the deep rooted nature of many accidents. The principles are easily applied to aviation;

Figure 3.1   Five levels of sociotechnical system

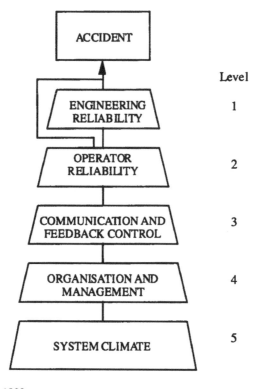

Source: Technica, 1992.

Level 1 Engineering Reliability The design of the hardware and software involved in operating aircraft and the limits within which it may operate. This includes airframes, engines, navigation aids and air traffic control. It does not include the man-machine interface, which is a function of the reliability of the operator. It does include automatic safety devices and systems designed to operate in emergencies such eg. ram air turbines. It is the easiest of the five components to quantify and the most commonly cited, in terms of failure rate or mean time to failure. The reliability of the aviation system is often measured in terms of engineering reliability for this reason.

Level 2 Operator Reliability This category encompasses all human factors that can influence the reliability of the operator such as training levels, experience, job design, workplace design and support systems. The amount of variation in operator reliability is a function of Levels 4 and 5, which set the standards for recruitment, training etc.

Level 3 Communication This level is concerned with the dissemination of and feedback from information through documentation, briefings, log books and reporting systems etc. It is the vital linkage between all levels of the sociotechnical system without which a level of coordination and safety is unachievable.

Level 4 Organisation and Management Style of management and the structure of an organisation will strongly affect the management of safety through the setting of standards, priorities and targets. Management must fit personnel and processes to meet the requirements of the system if it is to perform in the way they anticipate.

Level 5 System Climate The widest of the five levels linking the organisation and its management with other systems such as the regulator. Influences result from economic pressures, public opinion, regulations and the state of the technology available. The inclusion of this level is especially important for comparing how similar organisations (e.g. airlines) may operate in different parts of the world or under different types of regulation.

Understanding of the various levels of a system and their importance in terms of accident prevention diminishes inversely to the relative importance of each level in preventing a major accident in the future. For example, if the system climate is one of economic strife and poor regulation and management is more intent on *keeping aircraft flying* then the increasing of mean time to failure of, say, an engine component will have little affect on the overall safety of the system. If poorly trained crews are forced to fly an unfamiliar aircraft in marginal conditions then the extra reliability of such a component will probably have little or no bearing on the likelihood of them having an accident. However, if regulation is tight and economic pressure at an acceptable level then the increase in reliability of the component may not even be necessary to maintain a safe operation.

What is a Systems Approach?

For any accident to occur, there need to be a number of causal factors or conditions that combine in a particular way. No accidents occur solely as a result of a single factor. In aviation, accidents result from 'collective mistakes rather than individual errors' (Reason, 1992) where collective mistakes are the result of a number of individual errors interacting together. One way of representing an accident diagrammatically is by using a fault tree as follows:

Figure 3.2   Example fault tree

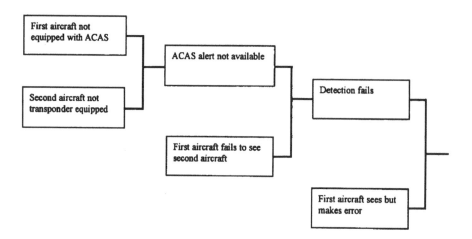

Source: Adapted from VRJ, 1994.

The fault tree shown in Figure 3.2 is part of a model of collision risk to aircraft under air traffic control.

Each box represents an error of omission (e.g. failure to fit ACAS equipment) or an error of commission (e.g. aircraft sees but makes error). Individual errors are not enough to cause an accident. The reader must be careful not to interpret errors late in the chain as being accident causing by themselves. For example, if a collision occurred following the error made by the first aircraft after spotting the second, traditional accident investigation may have deemed that the accident would not have happened if this error had not taken place and therefore this is the causal factor. Although the former statement bears some truth, the accident could have also been avoided through the fitment of ACAS alert systems or transponders. Further, the reasons behind the first crew's inability to successfully avoid the other aircraft, such as poor airmanship, may not have been significant if a collision pair had not been created.

Aviation represents a complex system of multiple interactions that is highly sensitive to both the physical environment and the passage of time. It requires the close interaction of both a technical and human component in a situation where both are highly dependent upon and significantly affected by each other.

James Reason's text *Human Error* (1990) has become a reference point for all human factors practitioners involved in safety. The underlying theme is that, '...in considering the human contribution to systems disasters, it is important to distinguish two types of error: Active errors, whose effects are felt almost immediately, and latent errors whose adverse consequences may lie dormant within the system for a long time, only becoming evident when they combine with others to breach the system's defences'.

In simple terms, Reason illustrates that accidents are the end result of a number of failures to the safety system. Therefore, if any given accident is assigned a single causal factor, it ignores the safety systems that had been built to prevent the last (failed) defence from ever needing to be used. Traditional accident investigation has tended to focus on the active failures that occurred immediately prior to the accident. These failures may include *pilot errors* such a failing to execute a stabilised approach, or controlled flight into terrain or other errors such as the failure of ATC to assure separation. However, in all of these cases, there are a number of safety systems (or at least opportunities for safety systems) which need to have failed prior to the event. These are what Reason refers to as *latent defects* and may include the failure of a crew member to cross-check, the failure of training staff to educate individuals correctly in how to deal with a particular situation, an economic decision not to fit a particular item of safety equipment and so on.

This is best described using what is now commonly referred to as Reason's *Swiss Cheese Diagram*. The barriers or walls represent the safety systems (or opportunities for safety systems) which all exhibit a number of holes or latent defects. This is a normal situation for most, if not all, complex systems and these defects are not enough to cause an accident by themselves. It is only when a number of latent defects *line up* and interact that at a particular moment, which combines with an active failure, that an accident occurs. Therefore it is possible for 9 out of 10 safety measures to fail or 999 out of a 1,000, both without an accident, but when the final *missing* failure occurs, so an accident will also occur.

Figure 3.3   Reason's organisational accident

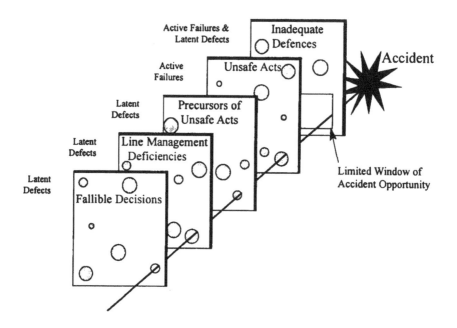

Source: Redrawn from Reason, 1990.

A clear example of such an accident occurred in 1987 when the Herald of Free Enterprise Ro-Ro ferry sank outside Zeebrugge harbour. The *blame* was initially placed upon the Assistant Bosun who was responsible for ensuring that the bow doors were closed for sailing. He was asleep in his cabin when the ship set sail and did not telephone the bridge to inform them that the door were still open on departure. Deeper examination of the accident demonstrates how the Assistant Bosun's actions were one of a few active failures or circumstances (eg. a swell in the North Sea) which had combined with a particular set of latent defects that had existed within the *Free Enterprise* class of vessel since they were commissioned.

The failure of the Assistant Bosun represented a *fail-unsafe* condition, which in turn is a function of poor system design. Other latent defects in the system previous to the incident include a management decision not to install a safety warning light (fallible decisions), poor crew scheduling, which had contributed to the Assistant Bosun's fatigue, (line management deficiencies) and the setting of precedence in allowing Ro-Ro ships to leave port with the bow doors open in spite of previous occurrences (psychological precursors of unsafe acts).

In 1988, The World Bank held a workshop '...to develop a multi-sectoral/multi-disciplinary research program to determine critical management and organisational failures that may lead to a catastrophic failure.' This provided a simple, yet effective diagram to demonstrate the inverse relationship between the level of understanding in accident prevention and the relative importance. The diagram is easily related to Technica's sociotechnical pyramid.

Figure 3.4   Sources of organisational failure

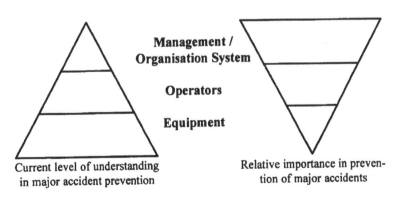

Management /
Organisation System

Operators

Equipment

Current level of understanding
in major accident prevention

Relative importance in prevention of major accidents

Source: Technica 1992.

As the World Bank pyramid clearly shows, the need for understanding is inversely proportional to the areas of current knowledge (namely the engineered and clearly defined world of automated safety devices and similar equipment). It is here that a systems approach to aviation safety is vital if it is to advance to any degree in the near future. A systems approach grasps the concept of accident chains and the existence of multiple causes in accidents. It acknowledges the potential of organisations and their culture have upon the *health* of a system and in doing so attempts to redress the imbalance.

Speaking at the 1997 Ohio State University Aviation Psychology Symposium in Columbus, Ohio, ICAO's Head of Human Factors, Captain Dan Maurino (1997) warned that '...human error is a symptom of system failure and not the cause of it.' This was a move aimed at challenging the new found temptation of the industry to shrug its collective shoulders and simply assign accidents to human error. The belief that human error is specific to individuals, such that following an incident or accident it is sufficient to just remove that individual from the system to regain safety, is both outdated and fundamentally flawed.

Knowledge of accidents and the multiple causes that lay behind all of them reveals that human error is present in 100% of cases (Faulkner, 1996). Such errors may reveal themselves as decisions far away from the accident face (such as in the certification process of a particular component or aircraft type) or may present themselves as active failures immediately prior to the event. There may be multiple human failures or a mixture of human and technical or environmental factors, but it is impossible for any accident involving a system as complex as aviation to be totally absent from an example of human error as a contributory factor. Once this fact is accepted, efforts can be better directed towards detecting human error (as a symptom of system failure) and preventing accidents by strengthening or adding defences.

The Use of a Systems Approach for this Case Study

The supposition that human error is present in 100% of aviation accidents is logical and supported by Reason's theory of organisational accidents. However, it is not a view that is generally held by the aviation community at large. This is for a number of reasons including insufficient information;

machismo or ego; incomplete accident investigation and a need to avoid blame / punishment / litigation.

However, the case-study design utilised in this research took account of the importance of human error. If it is present in 100% of accidents, then it seems logical that safe systems are such because of the control of human error. It is for this reason that the focus of this text is tilted particularly towards this area.

The understanding of all of the complex interactions of high technology systems is limited by what is termed *system opacity*. In other words, the high complexity of the system makes it difficult to understand how some actions may influence other areas of the system and in turn, this is often why accidents occur. Even moves aimed at increasing safety in a particular area can have a negative effect that is not immediately obvious.

A very simple example is that of domestic smoke alarms. Fire service data support the view that people who own smoke detectors are liable to become less cautious because they know they have a safety device. In many cases this has led to deaths where householders have allowed the alarm's battery to run flat and are then killed in a fire which burns undetected. In other words although the introduction of smoke alarms has undoubtedly saved many lives, they have also resulted in extra deaths through a new cause. As long as the latter figure is outweighed by the former and its set up costs then they will be seen as a useful device.

One of the reasons for looking at Australia's safety from a systems outlook is so that positive safety concepts which may be cancelled out elsewhere in the system are not overlooked. For example, if the weather is shown to be better than average for flying, it may be that more accidents occur in bad weather because of insufficient experience, Australian pilots are less able to cope with inclement weather overseas or pilots become blasé about the hazards. Although the contribution of good weather to the safety record would still stand, the negatives must also be presented to prevent the onset of complacency.

Review of Accident Types

Although accident investigation has started to move to a situation where accidents are recorded by multiple causal factors or contributory factors, statistics used to describe accident types generally still focus on a single primary cause. This is due in no small part to the fact that historical accident investigations have tended to settle for a single primary cause. In fact, the US National Transportation Safety Board is still required to determine the primary cause of an accident even if they do now include other significant causal factors.

While some of the more progressive investigative bodies such as Australia's ATSB and Canada's CTSB (Canadian Transportation Safety Board) now refuse to specify a primary cause in their accident reports, it will take a number of years before accident cause statistics start to reflect this way of thinking. In fact, even though these accidents are classified in the official accident report as having numerous causal factors, they are still graded into a single causal factor by insurance companies who produce their own sets of statistics. There are advantages for some of the publishers of accidents statistics to keep to the old system of single primary causes, not least the ease of compilation. An aircraft manufacturer, for example, may not appear much under the category of *aircraft design deficiencies* as a primary cause, but this may be a contributory cause to poor human performance, which would then be rated as the primary cause. If all contributory factors were included then the number of accidents where aircraft design deficiencies were a factor may be a much higher proportion.

The real danger from looking too closely at accident causation statistics (as with many forms of statistics) is taking them out of context to support or refute a particular argument. Each classification of factors attracts interest from different sectors of the industry and tends to be extensively reviewed at the micro-level. The objective of this section is to introduce the current statistics that are available in this area and the relative importance that they place upon different areas. In examining a particular operating system, this research had to take account of varying perceptions of the relative importance of individual causal factors or threats. For example, this perception can be a function of the amount of work done in a particular area rather than its relative importance.

Boeing's annual statistics (1996) separate primary cause factors into six distinct categories as shown in Figure 3.5. (The occurrence of acts of aggression and injury through turbulence and evacuation are removed.) The category of *flightcrew* is a somewhat controversial one, not least because it represents such a large component of the data. A senior airline official suggested that one '...should treat Boeing's statistics with a little caution'. The suggestion is that an aircraft manufacturer is unlikely to

advertise the fact that flightcrew errors that cause an accident are often partly the result of the aircraft design or equipment levels not compensating for
their actions.

Figure 3.5   Primary cause factors; all accidents (1959-1995)

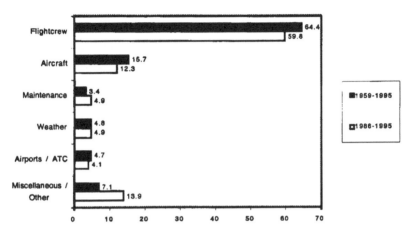

percentage of total accidents with known causes

Source: Redrawn from Boeing, 1996.

However, statistics produced by IATA (1993) demonstrate a similar picture
of the proportion of accident classification types. The classification system
used by IATA distinguishes between human, technical and environmental
factors. Therefore, Boeing's six categories would be matched as follows;

Boeing Classification	**IATA Classification**
Flightcrew	Human Factors
Airplane	Technical
Maintenance	Technical (and HF)
Weather	Environmental
Airport/ATC	Environmental (and HF)
Miscellaneous/other	Insufficient Information

For a ten year period (Boeing 1986-95; IATA 1984-91), IATA's human factors category actually accounted for 66.9% of accidents as oppose to Boeing's 59.8%. This is presumably a function of using the wider *human factors* category rather than the more specific *flightcrew* label which tends to suggest that human error is the exclusive domain of the aircraft's crew. The legend *human factors* will encompass maintenance and ATC errors. The technical category equated to 16.5% (Boeing = 17%) and the environmental category, 16.5% (Boeing = 9%). Assuming some variability for the different ten year periods used, small sample and different classification methods, the two sets of data appear to bear out similar conclusions about the types of accidents encountered. In particular, the high percentage of human factor primary cause accidents is significant as being the greatest challenge in aviation safety.

Table 3.1   Accident causal factors for 219 accidents

Group	Factor		No. of accidents where factor occurs
Aircraft systems	1.1	Non-fitment of currently available safety equipment (GPWS, TCAS, windshear warning)	47
	1.2	Failure / Inadequacy of safety equipment	3
	1.3	System failure - reduced controllability	18
	1.4	System failure - other	15
	1.5	Non-fitment of potential new equipment (enhanced GPWS)	31
ATC / Ground aids	2.1	Incorrect or inadequate instruction/advice	15
	2.2	Failure to provide separation	8
	2.3	Lack of ground aids	24
	2.4	Wake Turbulence - aircraft spacing	1
Atmospheric	3.1	Structural overload	1
	3.2	Wind shear / upset / turbulence	16
	3.3	Poor visibility	22
	3.4	Runway condition (ice, standing water etc.)	7
	3.5	Icing	5
Crew	4.1	Lack of situational awareness	75
	4.2	Incorrect selection on instrument / navaid	7
	4.3	Action on wrong control / instrument	4
	4.4	Slow / delayed action	23
	4.5	Omission of action / inappropriate action	57
	4.6	Failure to cross-check / co-ordinate	118

Group	Factor		No. of accidents where factor occurs
Crew (*cont.*)	4.7	Fatigue	6
	4.8	State of mind	5
	4.9	Fast / high on approach	9
	4.10	Slow / low on approach	13
	4.11	Loading error	4
	4.12	Flight handling	60
	4.13	Lack of qualification / training	11
	4.14	Incapacitation	2
	4.15	Failure to look-out	4
Engine	5.1	Engine failure	22
	5.2	Damage due to non-containment	5
Fire	6.1	Fire due to engine failure	4
	6.2	fire due to aircraft systems	4
	6.3	Fire - other causes	8
	6.4	Post crash fire	24
Maintenance	7.1	Failure to complete due maintenance	1
	7.2	Maintenance or repair error / inadequacy	12
	7.3	Ground staff struck by aircraft	2
Structure	8.1	Corrosion, fatigue	3
	8.2	Structural failure	7
Failings leading	9.1	Collision with high ground	49
to impact with	9.2	Collision with level ground / airport	35
terrain / obstacle	9.3	Impact with obstacle / obstruction	28
	9.4	Mid-air collision	8

Note: *Terrorism and sabotage, test and military-type operations are excluded. Also fatalities to third-parties who were not concerned with the operation of the aircraft are excluded.*

Source: Redrawn from Ashford, 1994.

Deeper examination of accidents gives a clearer picture of the significance of human factors in accident causes. Ashford (1994) examined 219 accidents (to aircraft above 5,700kg MTOW) documented within the UK CAA's World Aircraft Accident Summary and ICAO's ADREP database to highlight causal factors. The use of causal factors over primary causes is highly significant and separates this analysis from those of Boeing and IATA.

A quick glance at the chart demonstrates the prevalence of human factor problems in aircraft accidents. The 3 most common deficiencies are

4.6	Failure to cross-check / co-ordinate	(118 accidents)	(54%)
4.1	Lack of Situational Awareness	(75 accidents)	(34%)
4.12	Flight Handling	(60 accidents)	(27%)

However, it should be noted that such error types are combined with other failure types for accidents to occur. For example, lack of situational awareness may combine with the presence of high ground or the non-fitment of safety equipment such as GPWS or TCAS (i.e. lack of situational awareness by itself is not enough to cause an accident - neither is lack of GPWS or TCAS equipment).

The advantage of Ashford's multiple-cause approach is that it allows the researcher to work on multiple prevention approaches. In other words, one approach may eliminate a particular hazard at a cost of $n whereas another may eliminate two particular hazards at $2n. More research might then discover a third approach, which could counteract three hazards at only $n. The latter solution is the optimal choice but might not have been selected if individual causal factors were considered in isolation. For example, while collision with high ground (9.1) may be prevented through the use of GPWS equipment, it may also be prevented through improving situational awareness through human factors training (4.1), ensuring cross-checking of the flying pilot by other crew members through CRM training (4.6), the use of improved ground aids through infrastructure investment (2.3) and so on. In turn, by ensuring cross-checking of the flying pilot (4.6) this could also counteract the following accident causal factors;

4.2	Incorrect selection on instrument / navaid
4.3	Action on wrong control / instrument
4.4	Slow / delayed action
4.5	Omission of action / inappropriate action
4.9	Fast / high on approach
4.10	Slow / low on approach
4.12	Flight handling
4.14	Incapacitation
4.15	Failure to look-out.

Sears (1986) examined 93 major accidents from a selection of 126 that occurred between 1977-1984. In order to select only those causes that were significant, the following criteria were required to be met: Firstly that the accident might reasonably have been prevented had the factor not been present; and secondly, that a definitive solution or remedy can be envisioned for the elimination of the factor.

The product was a list of significant accident causal factors as they appeared in the 93 accidents. Of these, only 28% exhibited a single causal factor, 54% had two factors and 18% represented three or more causal factors. It is hard, if not impossible to believe that any accident involving a complex, high technology system such as aviation can be attributed to a single cause. It must therefore be assumed that Sears' data accurately reflects the product of an accident investigation process that has failed to document the complete set of causes. The list is presented in Table 3.2.

Table 3.2   Significant accident causes and their percentage of presence

33%	Pilot deviated from basic operational procedure
26%	Inadequate crosscheck by second crew member
13%	Design faults
12%	Maintenance and inspection deficiencies
10%	Complete absence of approach guidance
10%	Captain did not respond to crew input
9%	ATC failures or errors
9%	Crews not conditioned for proper response to abnormal conditions
9%	Other
8%	Weather information insufficient or in error
7%	Runway hazards
6%	ATC/crew communication deficiencies
6%	Pilot did not recognise the need for go-around
5%	No GPWS installed
5%	Weight or centre-of-gravity in error
4%	Deficiencies in accepted navigation procedures
4%	Pilot incapacitation

Source: Redrawn from Sears, 1986.

What Sears, Ashford, Boeing and IATA's analyses all have in common is the high proportion of human factor failures. Whether they are labelled as flightcrew causes or at least biased towards 'the last person to touch the controls', the fact is that the high proportion of human factors causes is even now underestimated. Although researchers appear to have settled on human error or pilot error (the terms are often misused interchangeably) as accounting for 60-70% of accidents, with even industry leaders such as Ohio State University Aviation's Richard Jensen (1982) stating that '...aviation accident statistics have found that 80-85% can be assigned broadly to pilot error...', there is a new movement gathering pace which supports the assertion that 100% of aircraft accidents are the result of some human error.

Students in the University of New South Wales' Aviation program (Faulkner, 1996) are challenged to find an accident that does not have a human factors contributory factor behind it. So far, no one has managed to rise to the challenge. This is because aviation is a systemic process, which is designed to operate within the known boundaries of the physical environment. It is only when operations exceed those known boundaries that accidents occur and the decision to exceed is always a human one; either consciously or subconsciously.

Once the aviation community accepts the human error component as being present in 100% of accidents, it is then able to move on to the systemic solutions rather than trying to find someone to blame. Complex systems will always be hampered by deficiencies in human performance whether these are errors or aspects of physical endurance. It is for this reason that the design of that system to be *human tolerant* is a priority.

By examining the accident and incident record, it is possible to get a reasonable idea about the areas that most need remedial action. However, existing knowledge of systems also issues its own warning, that in attempting to fix one problem, there is a risk of causing others. For example, whereas glass-cockpit technology has been able to reduce avionics failures and reduce workload, this has brought with it other problems such as reduced experience or reliance on automation, boredom and of training transfer from traditional style aircraft. Another example comes in the form of the increased accuracy of navigational systems. Whereas two aircraft that had been given opposing direction flight paths along the same route and at the same altitude may have narrowly missed as a result of instrument error in the past, more accurate technology has reduced this margin of error.

Figure 3.6   One of the problems of increased navaid accuracy

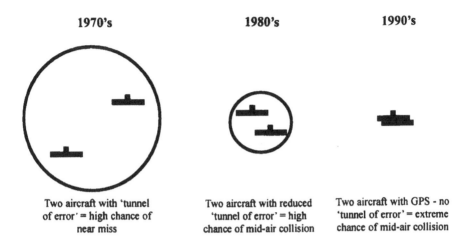

1970's	1980's	1990's
Two aircraft with 'tunnel of error' = high chance of near miss	Two aircraft with reduced 'tunnel of error' = high chance of mid-air collision	Two aircraft with GPS - no 'tunnel of error' = extreme chance of mid-air collision

There are no new accidents, only variations on a recurring theme. The use of a systemic approach to safety management is designed to prevent advances in one area from causing "old accidents" developing in new areas. Experience of accident investigation has demonstrated that (human) error was traditionally most important away from the accident face as latent defects within aircraft design, construction and flight planning. As these problems were confronted, accident investigation focussed more on the role of human performance in operation in the aircraft or air traffic control tower. Finally, investigation has entered into the new era of recognising the contribution of organisational (human) error, which supplements the other two areas.

The metamorphosis of Accident Causality

Date	Who / What was blamed?
Ancient Greece	Sun / Gods
Pre 1903	Gravity
Pre 1960s	Design / Structures
Pre 1990s	Pilot / Controller
1990s on	Organisational Failure

It is most important to note that although the causal factors behind many accidents are in fact the same, differences in perception caused by differences in thinking throughout the development of aviation have created a legacy of misinformation. The perception that many accident causes have been conquered brings with it a danger of complacency or at least an unintentional misdirection of efforts.

Early incidents involving the fly-by-wire A320 aircraft were swiftly blamed on the advanced technology when in fact the aircraft had behaved exactly as it was designed to. Poor type-conversion training resulted in handling accidents with the A320 the same way pilots had been caught out by the conversion from piston and turboprops to early jets such as the B707 and B727 series. One of the causal factors behind the infamous Mulhouse-Habsheim Airshow crash was the pilot attempting to execute a manoeuvre which was outside the capability of the aircraft which he had helped to program the fly-by-wire software for.

Why Look at Australian Aviation?

From the day a child is born, it learns from its mistakes. As it becomes older and wiser it will begin to learn not just from its own mistakes, but those of others to. As aviation developed in the early twentieth century, mistakes were one of the greatest sources of learning. Man's new found ability to fly was in an area where there were no experts and learning came at a price. The knowledge acquired through mishaps and accidents crafted the way aviation developed. For example, the frequency that control cables snapped led to their duplication and the problems of increased workload led to the advent of multiple crewing.

Accident prevention relied upon accident investigation and so the science of picking up the pieces and deducing what went wrong developed alongside aviation. Even as early as 1915, the Royal Flying Corps had appointed an *Inspector of Accidents* in the UK. In more recent years the skills of the aircraft accident investigator have ranked amongst the finest of their professions - engineers, forensic scientists, pathologists, sound engineers etc. When a terrorist bomb brought down Pan Am Flight 103 over Lockerbie in 1988, the wreckage trail was some 180 miles long. From this, the Air Accidents Investigation Branch was able to piece together not only the baggage container in which the device was planted, but also the suitcase and radio-cassette player it was hidden within. There are few unsolved large aircraft accidents, which pays great testimony to the ability of accident investigators.

However, just as the child becomes wiser by learning from other people's mistakes, so too the aviation industry needed to make this transition. Unfortunately, the accident record shows numerous examples of recurrent accidents before lessons were learned. Even now, the prevalence of Controlled Flight Into Terrain accidents (Between 1988 and 1993 for aircraft over 60,000lb, CFIT accounted for 53% of fatalities.) demonstrates how serviceable aircraft, even with state of the art equipment, are still being flown into the ground.

There are a number of key problems associated with relying upon accident investigation for advances in aviation safety:

a)   *Their reactive nature* By definition, accident investigation requires an accident to have taken place. In aviation, this often means expensive damage to equipment, third party property damage and serious injury, trauma or loss of life

Certain articles of ICAO Annex 13 mandate investigation of accidents. Each member state is required to investigate accidents that occur within their airspace or involve their aircraft and report the findings via ICAO. However, the depth of investigation varies considerably between states for various reasons including the interpretation of ICAO recommended practices by the regulator and the availability of resources for the investigation. As a result, accidents with similar causes may be investigated with differing outcomes or findings. In turn, this means that some of the opportunities for future prevention are lost and contributes to incomplete databases of accident causal factors.

All accident investigation authorities do not have the same level of resources available to them or indeed the same priorities as to what information is required from an accident. The latter disparity may range from the *who did it?* style of investigation to those that explore deeper organisational issues behind accident causes (for example, see BASI, 1996). Even within the leading investigative authorities such as the UK Accidents Investigation Branch (AAIB) and the Australian Transport Safety Bureau (ATSB) there are different priorities in investigation that are reflected in the staff positions. The AAIB, for example focuses on technical skills whereas ATSB has several human factors specialists.

Technically oriented investigations have been the traditional main-stay of aviation, discovering what went wrong on the day of the accident and highlighting a single primary cause. Recommendations promulgated from such investigations are usually concerned with technical engineering or operational aspects to remedy existing deficiencies. There is a general shift towards examining deeper human factors issues, but this seems to be a slow process.

b)   Their relative infrequency Although aircraft accidents are high pro-file events and are therefore perceived to be relatively frequent events, they are still relatively rare events. In the *jet-age* period 1959 - 1995, Boeing (1996) estimate a total of 304 million departures by western built aircraft (over 60,000lb MGTW) of which there were 1063 accidents.

Of these accidents, 850 were during passenger operations, 120 dur-ing all-cargo flights and 93 during test, training, demonstration or ferry. The 1,063 accidents can also be split between nine manufacturers which covers 32 significant types of aircraft. (i.e. this does not even split by series so a B737-100 would be classified alongside the more advanced -400 series.)

Add to this mix the issues mentioned above regarding the inves-tigative approaches of different states and the opportunities for trend analy-sis of accident causal factors become quite limited.

c)   The effect of time Strongly related to the above category is the pas-sage of time and its effect on the value of accident investigation results for the promotion of flight safety. The infrequency of accidents dictates that the compilation of any sort of trend statistics will cover a significant period of time. During that time period, it is reasonable to assume that a number of changes will have occurred in some or all of the following areas:

1.   Aircraft technology;
        new aircraft types
        new aircraft series
        existing aircraft modification
        retrofitted equipment

2.   Crew training;
                airline training methods
                training technology
                background of recruits
                use of on-condition monitoring

3.   Air traffic services;
                training of controllers
                new technology (e.g radar coverage)

4.   Accident investigation;
                new technology (e.g. flight data recorders)
                change of focus (e.g. systemic investigations).

Although there are some negatives to accident investigation, there are alternatives available which are more proactive in nature;

i. Incident Investigation Accidents represent the catastrophic breakdown of any safety system and yet there are many more occasions where safety systems fail either partially or without serious consequence. Such events are termed as incidents and are significantly more prevalent than accidents. The analysis of incidents may initially seem to provide similar opportunities for the prevention to the investigation of accidents, as the causal factors are often similar. Indeed, Pidgeon (1991) observes that near misses differ from disasters '...only by the absence of the final trigger event and the intervention of chance.' While this may be the case, the nature of incidents mean that the reporting mechanisms aren't anywhere near as effective as they are for accidents. Incident reporting tends to be voluntary and depend upon individuals not only taking the time and effort to report incidents but also recognising incidents in the first place. For example, although failure to cross-check or loss of situational awareness are found to be present in a high percentage of accidents, they are rarely reported in incident reporting systems because either they are not detected or the crew member involved is either too embarrassed, or in fear of punishment to report it.

    If a complete profile of incidents could be collected then the usefulness of this type of data would be significantly enhanced. Improvements in reporting forms and feedback processes have helped to improve quality and response rate although, as many systems are run at company level and are confidential, databases tend to be somewhat fragmented.

However, as the various databases of incidents grow, so the opportunities to learn more from trend analysis will develop. However, this will take some time and will require some caution. Just because a particular reported incident type appears to represent a high proportion, does not mean that it is in fact the most serious problem. At its very simplest, the fact that a large proportion of individuals recognise and report a problem type may mean that they are already modifying their behaviour to take account of it. Conversely, this may also mean that the poorly reported incident types are the ones that are of the greatest concern to the safety profession. However, this is not to say that there is a pareto relationship and will need further research.

At a higher level, there is significant potential for accident investigative techniques to be applied to incidents on a larger scale than simply collecting voluntary incident reports. At present, few authorities have the resources available to investigate incidents as well as accidents although there is a laudable move towards looking at some major incidents. For example, the Australian BASI investigated a significant near miss at Sydney Airport when a DC10 and A320 nearly collided during simultaneous intersecting runway SIMOPS operations. There are also procedures in many organisations to conduct their own investigations, although by their very nature, the results of the investigations tend to remain exclusively in-house and not available to the research community.

ii. Case Study Research If accident investigations can be seen to be an in-depth case study of a failure to a safety system, it seems most reasonable that *safe* normal operations provide a case study of a successful safety system. As mentioned earlier, Reason (1993) suggested; 'Should we not be studying what makes organisations relatively safe rather than focussing upon their moments of unsafety? Would it not be a good idea to identify the safest carrier, the most reliable maintainer and the best ATC system and then try to find out what makes them good and whether or not these ingredients could be bottled and handed on?'.

By examining healthy operations, researchers are not constrained by the problems that often occur following accidents. (Such as injured or dead witnesses and those who wish to cover up their mistakes.) In extreme circumstances, accidents remain unsolved in the absence of witnesses (such as the Colorado Springs and Pittsburg B737 accidents) or are prone to the distortion of facts that may or may not be discovered. For example, in the case of the 1979 Mount Erebus Disaster, the subsequent Royal Commission (Mahon, 1981) was forced to make the following damning conclusion about the airline management's contribution to the inquiry;

> No judicial officer ever wishes to be compelled to say that he has listened to evidence which is false. He prefers to say, as I hope the hundreds of judgements which I have written will illustrate, that he cannot accept the relevant explanation, or he prefers a contrary version set out in the evidence.
>
> But in this case, the palpably false sections of evidence which I heard could not be the result of a mistake, or faulty recollection. They originated, I am compelled to say, in a pre-determined plan of deception. They were clearly part of an attempt to conceal a series of disastrous administative blunders and so, in regard to particular items of evidence to which I have referred, I am forced reluctantly to say that I have had to listen to an orchestrated litany of lies.

Mahon's statement is unusual, but only because investigations or inquiries of the magnitude of the Royal Commission into the Erebus Disaster are few and far between. The new found understanding of organisational accidents afforded by the work of academics such as James Reason has shown that high level organisational failures of this nature are commonplace in high-technology industry accidents. This includes rail, maritime and space transportation (e.g. Clapham Junction, Herald of Free Enterprise and Challenger Disasters), chemical, nuclear power and oil production (Union Carbide at Bhopal, Three Mile Island and Piper Alpha in the North Sea).

The use of *good examples* increases the willingness of expert witnesses to provide information and represents a more positive approach than a critique of failures and errors. The benefits fall into two major categories - namely in terms of enhancing the safety of other operations through example and secondly, in aiding in the preservation of the original safe operation.

4 Exploring the Physical Geography

The Natural Environment

Anecdotal evidence regarding the reasons behind Australia's apparently good safety record often targets three categories, namely those of good weather, low traffic density and terrain. When flight crews of large-jet aircraft in Australia were asked 'How do you think that Australia has managed to achieve the record of zero hull losses for jet RPT operations?', the top factor, mentioned by approximately 58% of respondents was the weather with low traffic density in 3rd place (33%) and terrain in 6th place. (28%)

Figure 4.1   **What has kept Australian aviation safe?**
                   (Answers from Australian large-jet transport pilots)

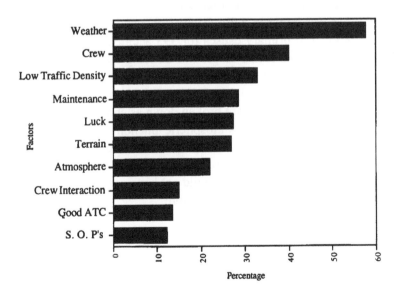

The next two chapters examine the influence of the natural environment upon aviation safety in Australian. The first examines the subject of physical geography and the second, the effect of distance on the development of the aviation system.

Consideration is also given to the fact that Qantas and more recently, Ansett Australia, also operate outside of Australia. By definition half of all take-offs and landings experienced by international flights will occur outside of the country and this risk must be taken into account when examining the carriers' safety records.

Aviation Meteorology

The principles of flight are based on good weather; clear air, light wind and no precipitation. As soon as weather phenomena are added to the equation, the ability to fly becomes a compromise.

Many believe that Australia benefits from good aviation weather. However, the challenge for the safety investigator is to discover whether this view is a major factor behind the safety record.

In terms of primary cause, the percentage of hull loss accidents due to weather phenomena is relatively low. According to Boeing figures (1993), the average percentage of total accidents with known primary cause due to weather was between 4.9% (1959-1992) and 3.3% (1983-1992). A further breakdown of primary causes for 393 accidents between 1982 and 1991 attributes nine accidents to ground de-icing / anti-icing and ten to windshear, with a total loss of 765 souls. Ashford's (1994) analysis of 219 large aircraft accidents (see Table 2.1) reveal atmospheric factors were causal factors 51 times. These are broken down as show in Table 4.1.

The disparity between atmospheric phenomena as causal factors and primary causes is attributable to the fact that aircraft and ground aids are designed, and crew trained to operate in adverse weather. Many accidents involving weather as a factor would not have occurred without a compounding equipment or human factor failure.

Table 4.1   Breakdown of weather-related causal factors

Factor	No. accidents where factor occurs
Structural overload	1
Wind shear / upset / turbulence	16
Poor visibility	22
Runway condition (ice, slippery, standing water etc.)	7
Icing	5

Source: Adapted from Ashford, 1994.

Weather is a function of physical geography and is affected by climate, relief, latitude and season. It may be considered to be extraneous to the man-made aviation system and hence largely uncontrollable. However, there are very few weather phenomena that cannot be compensated for, and those that remain, such as low level windshear, are obviously the most hazardous. In addition, the geographic distribution of most weather phenomena is well known which means that equipment and crew training can accommodate expected hazards. The accident record demonstrates a small number of accidents where nature's forces have been stronger than predicted and jet aircraft have just broken up in flight.

Weather Types and their Occurrence in Australia

In Australia, the Bureau of Meteorology issue warnings about a number of weather phenomena, which are significant to aviation safety.

Hailstones Hailstones form when raindrops freeze at high levels. They grow as they are recycled through powerful up- and down-draughts. Hailstones the size of cricket balls have been recorded in Australia, which can cause severe damage to aircraft. In April 1999 a major hailstorm caused millions of dollars of damage in Sydney including a large number of aircraft at the airport. Hail can cause contamination of runway surfaces and can seriously obstruct visibility. In 1977, an American DC-9 crashed onto a State highway after giant hailstones cracked its windshield and its two engines had flamed out, with the loss of 62 souls. It is the up- and down-draughts of wind (which enlarge hail) that are, perhaps, an even greater threat as they are not always associated with hail but are always dangerous to aircraft in the form of dry or wet windshear.

Windgusts Falling rain and hail drag the surrounding air downwards, accelerated by evaporation of raindrops cooling adjacent air. Upon reaching the ground, the air spread as an outflow front, which may even curl to form a horizontal vortex. Windgusts may cause aircraft to land offside the runway if on approach it encounters a crosswind. A sudden change in wind direction during the critical phase of landing can seriously affect landing speed and even cause the aircraft to stall. If an aircraft flies through the centre of a storm cell, the wind direction may suddenly change from a strong headwind (good for landing) to a strong tailwind (poor for landing) which may force the aircraft into the ground.

On 2nd August, 1985, a Delta L-1011 was on final approach to Dallas-Fort Worth Airport through the middle of a storm. As it passed through the core of the storm, the wind speed suddenly altered as a consequence of a microburst and associated windshear. The aircraft crashed on a highway, 1,800m short of the runway, with a loss of 135 souls.

Heavy Rainfall Heavy rainfall poses a two-fold threat to aircraft. The first is that in flight, rain can cause engines to flame out as the combustion chamber ingests too much water and too little oxygen. Less rare is the effect of rain at airport terminal level, where it can seriously affect visibility and can contaminate the runway. Standing water can dramatically reduce the coefficient of friction of runway pavements to zero, when aircraft begin to aquaplane. Aquaplaning is a particularly hazardous condition as the aircraft can neither make adhesion to brake, nor can it gain traction to execute a go-around. Aquaplaning is both a function of the intensity of rainfall, tyre pressure and the type and condition of the pavement.

The amount and intensity of rainfall in Australia is often underestimated because it tends to fall in short, heavy bursts and often occurs at night. In fact, Sydney actually boasts a higher annual rainfall than London. Rain falls at Sydney airport on an average of between 11 and 14 days per month with mean monthly rainfall of between 69 and 135mm. (BOM, 1989). The highest recorded rainfall for a month was 643mm in June 1950 and the highest amount on one day was recorded at 281mm in March 1942. At the other extreme, Darwin, in the tropical belt, experiences between zero and 21 days of rainfall on average, per month. Mean monthly rainfall ranges from 1 to 409mm with the highest recorded monthly rainfall occurring in March 1977 (1,104mm) and the highest daily rainfall in January 1897 (296mm).

Rainfall is generally not a serious problem to RPT level aviation by itself, but can often be compounded through a series of other factors. For example, when associated with strong downdraughts in the form of microbursts, or contributing to poor visibility on marginal approaches. As Ashford (1994) notes, the highest proportion of weather contributory factors relate to poor visibility which can lead to unstable approaches.

Tornadoes These are a rare and very violent product of thunderstorms, and consist of a rapidly rotating column of air in a funnel shape. The vortex of a tornado can range in width from a few metres to several hundred metres and may contain winds that reach 450 km/h. They are infrequent, easily visible on radar and therefore avoidable. Whilst they pose an obvious threat to civil aircraft, they have not been responsible for any jet RPT aircraft losses.

Mountain Waves Mountain waves occur when relief such as mountains or escarpments affects wind speed and pressure. In much the same way that the shape of a wing creates lift, sharp topographical features can create clear air turbulence (CAT). Sometimes known as *Karman Vortices*, they are also known to occur around large structures such as power station cooling towers and skyscrapers.

In Australia, the Bureau of Meteorology highlights a number of areas where the topography lends itself to mountain waves and CAT. These include Perth, Adelaide, Richmond, Badgery's Creek and to a lesser degree at island airports such as Lord Howe and Great Keppel. Mountain waves are difficult to see, especially if they occur in clear air, although the presence of lenticular cloud in the lee of mountains are good indicators of waves or severe turbulence. As they form due to topographical features, they are more of a problem for general aviation aircraft, and larger aircraft on climb and approach. Areas where such phenomena occur are generally predictable and therefore avoidable, although there is never room for complacency, even for large aircraft. BOAC lost a B707 near Mount Fuji, Japan due to the effect of strong lee waves.

Sand / Dust Storms Dust storms usually occur in late spring or summer but are relatively infrequent except in times of severe drought. They affect visibility over a wide area and can be potentially damaging to aircraft engines. They are associated with periods of intensive heating, fresh winds and down draughts from thunderstorms. There are have been no major accidents attributed to this type of weather phenomena, but it can be a contributory factor to other active failures such as loss of visibility on approach or dry windshear.

Icing One of the most severe weather hazards to the safe operation of aircraft is that of icing. There are two separate types of icing hazards, namely airframe icing and engine icing:

   a) Airframe icing This term is used where deposits of ice adhere to the structure of an aircraft where the outside air temperature is at or less than 0°C. Airframe icing is not confined to aircraft in flight and may actually occur on the ground, especially on the cold metal of an aircraft's structure. Icing is also not confined to visual precipitation conditions and can thus be associated with clear air.

   There are three types of ice that may occur on an aircraft, namely hoar frost, rime ice and clear ice. Hoar frost is a crystalline deposit formed in clear air where temperature is less than 0°C and relative humidity is high. It can occur to aircraft on the ground as frost, which is not particularly hazardous, provided it is removed before take-off. Hoar frost can affect the aerodynamic efficiency of an aircraft's lift surfaces and reduce vision.

   Rime is a white, lumpy deposit formed by the rapid freezing of supercooled water droplets that contact the airframe. The supercooling traps air between the ice crystals which creates an opacity that makes rime ice easily visible. It is most common on leading edges and intakes, and can form between 0°C and -40°C, but is usually encountered between -10°C and -20°C. It is most common in stratiform cloud but may also be encountered in cumuliform. As well as disrupting the airflow over an airframe, rime ice can also block air intakes and pitot tubes.

   Clear or glaze ice is a transparent sheet of ice that can be smooth or rippled in its form. It is usually formed by the slow freezing of supercooled water droplets, especially on the ground and in temperatures between 0°C and -15°C. Glaze ice adheres very strongly to the airframe and is therefore the most dangerous type of airframe icing.

In 1990, a SAS MD-80 crashed soon after take off as clear ice was ingested into the fuselage mounted engines. The aircraft had been incorrectly ground de-iced, partly because of the almost invisible nature of clear ice. In flight, the most hazardous forms of ice develop when the aircraft flies through supercooled rain or drizzle as the aircraft may suddenly become covered in a layer of clear ice.

The formation of airframe icing is affected by a number of factors such as air temperature and moisture. At altitudes where the temperature is less than -41°C, water clouds are rare and therefore the risk of icing is minimal. Heavy icing is also uncommon below -25°C but above this, the risk increases as the temperature raises to 0°C, especially for clear ice. Clouds with a high water content or large water droplets (such as cumuliform) tend to produce clear ice although aircraft do not tend to fly through long periods of such cloud unless flying along a front. The relief of overflown terrain can increase the depth of cloud and the size of water droplets so icing can be expected to be more severe in elevated areas.

Ground icing within Australia is not a major problem because of the climate and location of major airports. Only a very small part of the landmass ever experiences average daily minimum temperatures below 0°C. However, this is not to say that ground temperatures do not drop below freezing. Figure 4.2 plots contours of frost days (where temperature is less than or equal to 2°C) for Australia and initially seems to show a fairly high occurrence of frost. However, close examination reveals that high volume traffic airports such as Sydney, Brisbane, Perth, Darwin and Cairns lie close to, or on, the zero contour. Melbourne, however, does lie on the 50 day contour and Canberra, Launceston and Hobart all lie on the 200 contour. De-icing facilities are available at all of the frost-prone major ports although jet traffic from airports such as Canberra is relatively light.

b) Engine icing This type of icing can affect different types of engines in different ways. Carburettor engines suffer icing from the air intake, but as these types of engine are not used for Commercial RPT operations, they will not be considered within this text. For gas turbine (turbofan) engines, icing can affect the fuel systems and air inlets, which can in extreme conditions cause a flame out or compressor blade damage.

An in-flight emergency involving power loss on all four engines of a BAe146 occurred in 1992 when the aircraft entered unusually warm air at high altitude (31,000 ft) over Western Australia. 'The investigation determined that during high altitude cruise, the aircraft entered an area of moist air significantly warmer than the surrounding air. This resulted in a need to select engine and airframe anti-ice which in turn placed high bleed air demand on the engines. Under these conditions the fuel control units were unable to schedule sufficient fuel to the engines, thereby causing them to lose power, a phenomenon known as *roll-back*.' (BASI, 1994) This incident highlighted an operational limitation of the BAe146 series and has been compensated for by restricting the operating ceiling of the aircraft in Australia.

Figure 4.2   Australian frost period (days per year)

Source: Climate of Australia (1989) Bureau of Meteorology. Commonwealth of Australia copyright reproduced by permission.

The loss of the Seaview Commander 690B en route to Lord Howe Island was not a commercial jet RPT operation. However, the BASI (1996) findings indicate that there were both icing conditions at the time of the aircraft's disappearance and modifications for flying in icing that had not been done, which were contributory factors to the accident. Australia is not immune to in-flight icing conditions and there is no room for complacency.

Volcanic Ash   Volcanic ash can reach high altitudes and can be extremely hot and large in particle size. Simkin (1994) explains that although volcano belts cover less than 0.6% of the earth's surface, at least 1,300 have erupted in the last 10,000 years. Also, as volcanoes tend to have long active lives, it is very likely they will erupt again in the future although typically only at a rate of 60 per annum. During the years 1975-85, Simkin notes that more than 63 eruption penetrated aircraft cruising altitudes. Of the 16 largest eruptions over the last two centuries, all but four have been the first known from that volcano which emphasises the need for up to date information to keep aircraft separated from eruptions.

In 1982, a British Airways B747 overflew Mount Galunggung, Java and suffered multiple engine failure. It had unwittingly flown through the centre of a volcanic ash cloud. Although the aircraft eventually managed to recover altitude, the ash had fused itself to most of the airframe and scoured all of the windows. A weather satellite had photographed the eruption, but the information had not been passed on to the airlines. On-board weather radar is incapable of detecting such conditions as it reflects off water droplets, not dry ash.

There are no active volcanoes in Australia although volcanic dust clouds do drift into Australian airspace from the north. However, there is a major string of active volcanoes to the north of Australia that sit below major air routes. The main hazard to intercontinental flights is at night, especially from Plinian eruptions that eject large amounts of fine ash and gas to high altitudes.

The meteorological authority responsible for the flight information region issues SIGMET (Significant Meteorological Advisory) notices in question. They will be the result of an eruption notification and confirmation from the Japanese geostationary meteorological satellite (GMS) which is updated hourly. The system relies upon notification of any initial or subsequent eruptions from the national responsible body of the volcano's origin. The GMS can track the progress of an ash cloud although water/ice clouds can make ash cloud difficult to distinguish. According to Whitby and Potts (1994), during 1985-90, a total of 81 volcanic ash SIGMETs (or VOLMETs) were issued and in 80% of these it was not possible to identify and track ash clouds because of the presence of water/ice clouds. Further to this there is little knowledge of how long an ash cloud remains a hazard to aviation.

Australia experiences a variety of extreme weather conditions, not least because of its magnitude and the fact that it covers several climatic zones. However, the general perception remains that the country is blessed with good aviation weather. '...Weather tends to be stable for most of the year' and, according to Smith (1993), '...ice and fog are virtually unheard of'. This means that not only are aircraft less exposed to variable weather conditions, but they are forced to fly in marginal conditions for less time.

However, this may also mean that aircrews have little experience of flying in poor weather and consequently may be more likely to make significant errors in such conditions. It also does not provide an explanation for the fact that Qantas have never had a major weather related accident at an overseas port. (Note that discussion regarding the 1999 *QF1* accident and Bangkok is contained later in the text)

It is fair to say that, on average, Australian airports do not suffer from the worst types of marginal conditions i.e. standing water, slush, snow or ice, especially when compared to European or North American locations. Indeed, a survey of main Australian airports seemed to support this concept: Jack Caine, Duty Manager for Perth Airport reports that, 'Very few hours have been lost in Perth due to bad weather; there wouldn't be more than five hours closure in the last ten years.' (Caine, 1994). This is a similar story to Brisbane where Hall (1994) states that, 'Hours closed to bad weather in last ten years would be approximately 4-6 hours per annum. Frequency of fog cover approximately 2-3 hours, 4-6 times per year.'

However, this is a potentially incomplete picture, which needs some balance to account for *a lack of incidence verses a lack of incidents* and to highlight the potential problems of occasionally bad weather.

A lack of accidents due to a particular cause is not necessarily indicative of an absence of the hazard. In fact, the relationship may be exactly the opposite; an acute awareness of a particular hazard, such as high terrain, may be the reason that a lack of accidents has occurred. Hence, although a high number of weather related accidents in a certain area may be a function of geography, a low number of weather related incidents in a particular area may actually say nothing about the level of incidence of that particular phenomena. In other words, just because there have been few major accidents or incidents in Australia where weather has been a major contributory factor, does not mean that the weather is unquestionably good. Indeed, the truth might be as simple as a lack of traffic or even a lack of reporting. For example, windshear accidents involving GA aircraft may go undetermined if the pilots is killed on account of a lack of CVR / DFDR equipment and the short lived nature of shears.

Microbursts and Windshear

Windshear related accidents and incidents pose a significant threat to the safe operation of all sizes of aircraft. A US National Academy of Sciences report (National Research Council, 1983) documents 27 windshear events involving aircraft over 12,500lbs in the 18 year period between 1964 and 1982 with a total loss of 491 souls. The FAA's Advisory Circular 00-54 (FAA, 1988) also documents a total of 51 windshear related events between 1959 and 1983.

The Joint Airport Weather Studies Project (McCarthy and Serafin, 1984) adds that at least 27 civil airline windshear / microburst accidents and incidents had occurred between 1964 and 1984 resulting in 491 fatalities and 206 serious injuries. They also suggest that although the concept of microbursts dates back to the work of Fujita (1978) there were a number of accidents prior to that time that were probably the result of windshear (including an F-27 at Bathurst, NSW in 1974).

Windshear is a phenomena which regularly effects flying operations. However, severe windshear is recognised as '...a serious hazard to airplanes during takeoff and approach' (FAA, 1988). Melvin (1994) adds that 'Too many windshear accidents have been analysed with an emphasis on pilot error... In most cases the analyses were flawed... and this has caused considerable misunderstanding of various aspects of windshear hazards.' In Australia, Terrell (1988) adds the warning that '...the detection of severe windshear in the terminal area (particularly microburst activity within thunderstorms) is presently unsatisfactory.'

The effect of severe windshear, especially at low levels, appears initially to be most prolific in the USA. This is not to suggest a direct causal effect of the geography of North America. As Potts (1991) notes

Few aircraft encounters attributable to downburst encounters have occurred outside the continental USA, largely because of the lower traffic densities in those areas of the globe conducive to their development.

Due to a lower density of aviation traffic and more conservative air traffic control and pilot practices, the threat posed by microbursts in Australia is less severe than in the USA. This is reflected in the very low number of accidents in Australia attributed to downburst or microburst encounters. Notwithstanding this, it is evident that microburst pose a threat over tropical Australia which has previously been unrecognised and previous studies suggest there is also a significant hazard in southern Australia. As traffic density increases at airports around Australia the likelihood of a microburst encounter will also increase.

'When one considers that "moderate" rain echoes would not require the Australian Terminal Area Severe Turbulence (TAST) system to be activated, together with the observation that annual thunderstorm occurrence at Sydney (~28) and Brisbane (~35) are comparable to Chicago (~34), a significant hazard to safe operation is considered to presently exist and such hazard will increase with traffic density.' (Spillane and Lourensz, 1986)

The research question was therefore; how can it be determined whether continental USA has a greater incidence of microburst than mainland Australia? David Hinton of the NASA Langley Research Centre is acknowledged to be one of the world's leading experts in the occurrence of hazardous windshears and their effect on aircraft. (Bracalente, 1995) Hinton suggests that the only certain method of answering this question is not practical. In the absence of any previous national study in Australia, this would mean 'field studies requiring installation of large doppler radars... over a period of years to determine the frequency near major airports.' Middleton (1996) adds that although there is a doppler radar at Sydney Kingsford Smith Airport it is placed in a position that is sheltered from this type of occurrence and no useful research data is currently collected.

The only other method suggested involves estimating from the number of days per year of weather producing high microburst potential. Although the FAA's Advisory Circular AC00-54 (FAA, 1988) attempted to map microburst potential, there is some question as to its universal validity. The apparent low level of thunderstorm days over Australia, for example, is in marked contrast to the findings of the two studies that have been conducted there. Melvin (1994) observes that '...no evidence exists that any of the known microburst encounters have occurred in supercell storm cells' even though '...many pilots have been trained to avoid these in the belief that this will prevent any encounters.' Fujita and Caracena, recognised authorities in this field, have repeatedly emphasised that '...microbursts are frequently generated from benign-appearing cells.' As the FAA map did not differentiate between severity of storms it is difficult to accept the contours as reliable indicators of the risk of severe microburst encounters.

The Australian Bureau of Meteorology has not attempted to map the frequency of microburst / windshear encounters as such, but were able to produce a graph that helps to predict where these events are most likely. Produced from the Bureau's records of severe thunderstorm activity, they were able to plot the following contours, as shown in Figure 4.4.

Figure 4.3   Annual thunderstorm frequency

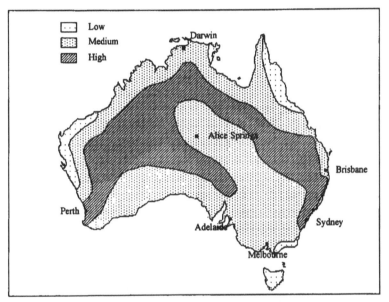

Source: Adapted from Bureau of Meteorology, 1995.

The Bureau adds that 'The geographical spread of severe thunderstorms in Australia is difficult to determine because of our low population density and lack of observations over most of the continent. While records of storm impacts show that the most damaging storms have occurred in the populous south east quarter of the continent, analysis of wind, hail and tornado data suggests that severe thunderstorms are a significant threat throughout the Country.' (BOM, 1996)

Of note is that once the levels of aviation activity are added to the above map, it becomes especially pertinent that Sydney, Perth and the Gold

Coast are all contained within the high storm zone. Darwin on the other hand, where anecdotally, the most severe storms are, is only rated as moderate. In 1991, Sydney airport saw 235,400 aircraft movements, nearly double that of its nearest competitor in Australia, Melbourne (134,900 movements). The position of Sydney Kingsford Smith Airport next to the coast means that when storms occur, they are usually severe in nature. In many cases they will hit aircraft that are either taking off heavily laden for, or arriving from long sector international flights. Add to this the fact that overseas pilots may rarely make landings at Sydney, will be subject to traffic restrictions and changing arrival procedures and the latent conditions that need to exist prior to a crash appear to be in place.

Crew Fears Regarding Weather

Although one of the most frequently cited anecdotal reasons behind Australia's apparently good aviation safety record is the *blue sky* or benign weather theory, the question of windshear / microburst encounters are very much the concern of Australian flight crews. Pilots were all asked to rank *the top three factors they considered posed the greatest threat to their flying safety*. The result was a little unexpected, particularly in respect to the answers given as to why Australia had managed to stay safe (figure 4.4).

The Qantas crews, which were a mixture of domestic and international, rated the threat of mid-air collision to be their greatest fear. However, a very close second, chosen by nearly 17% as one of their top three fears, was that of windshear / microburst encounters. In both cases, the type of incident is one where the flight crew will have little control over the outcome if the event occurred. Mid-air collision involving a large jet aircraft and any other type of aircraft are generally fatal because of the energy of impact. Microburst / Windshear encounters which occur at low level on take off or landing tend to leave the pilots of modern jets with a technological impossibility. If the pilot has reduced thrust to avoid overspeed when the aircraft first hits a microburst and then attempts to re-apply thrust he must wait for the engines to spool up which usually takes more time than is afforded by the altitude involved and the strength of the down-draught.

**Figure 4.4   What are the top 3 threats to your flying safety?
(Answers for Qantas Airways)**

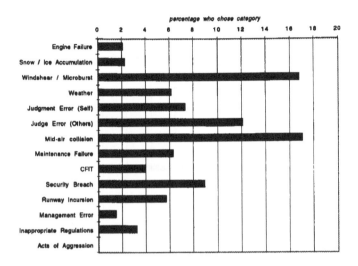

The results were then split to represent Domestic and International crews to examine whether the windshear / microburst threat was perceived to be greater for those flights which operated out of Australia. The two graphs are presented in Figures 4.5 and 4.6. The difference between Domestic and International crew results is negligible for the category *microburst / windshear* which suggests that they do not consider the problem to be a function of geography. The international crews do rate that factor over and above anything else, unlike the domestic sector which rates mid-air collision to be slightly more of a problem. At the time, only the International aircraft were fitted with Traffic Alert and Collision Avoidance System (TCAS).

**Figure 4.5   What are the top 3 threats to your flying safety?
(Answers for Qantas Airways International: B747)**

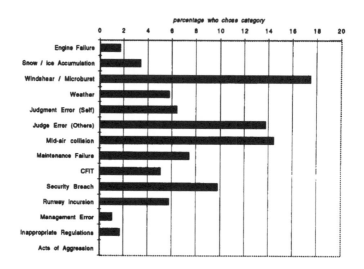

**Figure 4.6   What are the top 3 threats to your flying safety?
(Answers for Qantas Airways Domestic: B737, A300)**

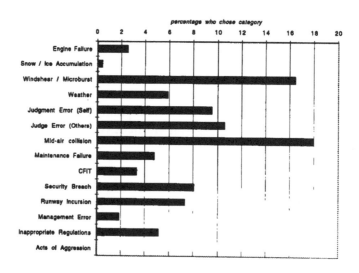

The results from Ansett Australia (Figure 4.7) show a significant difference between their opinions and that of Qantas crews, particularly in the area of mid-air collision risk and the split of judgment error between themselves and others. The importance of microburst / windshear encounters is only heightened by this response in that it is very much the prime concern of crew members. This diminishes the possibility that the Qantas result has been skewed by the experience of those pilots flying international sectors. The risk in Australia is not reasonably proven to be any less than anywhere else in the world and even if the prevalence of this weather phenomena could be shown to be significantly less, this opens up the possibility that infrequent exposure can mean low experience.

Figure 4.7   What are the top 3 threats to your flying safety? (Answers for Ansett Australia)

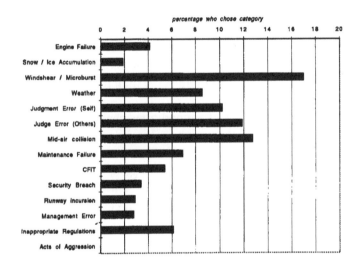

Predictive windshear detecting radar was approved by the FAA in 1995 but by 2000 neither Ansett Australia nor Qantas have announced any intention of purchasing such a system.

Weather and Crew Perceptions

Whilst weather conditions may be shown to be relatively less extreme than elsewhere in the world, the fact only counts for something if operations reflect that additional margin of safety. Risk homeostasis theory might suggest that faced with severe weather conditions, pilots will modify their performance to a *safe* level and therefore in good weather conditions, a similar modification of behaviour occurs. The argument to be answered is whether pilots who have little experience of flying in poor weather conditions compensate for this through added caution. Conversely, do pilots who are used to flying in poor weather conditions fly better because they have a lot of experience or is this cancelled out by behaviour modification?

Analysis of weather related accidents re-iterates the importance of the human component as the single greatest factor in aviation safety. Although there are still some areas open to debate, aviation meteorology is a relatively well developed science and many *weather* accidents occur because of fallible decision making on the part of the crew or other humans in the system. Such decisions can range from the *press on* attitude, where not all factors are considered in making a risk-taking decision, to high level policy decisions regarding, for example, the fitment of weather or predictive windshear radar.

Attitudes towards weather conditions can be just as critical as the actual weather conditions themselves. This is particularly true for high-capacity RPT aircraft, which are more able to cope with extreme weather conditions than GA aircraft. The decision to go or not go can be influenced by many considerations, some of which are non-operational. Subtle commercial or organisational pressures can sully the judgment of even the most competent crews as the accident record shows.

On 10th March 1989, an Air Ontario Fokker F-28 failed to climb on take-off from Dryden, crashing approximately 1km beyond the end of the runway. The aircraft's wings were contaminated with snow and ice following refuelling and a delay after ground de-icing. The investigation into the accident was far reaching (See Moshansky, 1992) and examined a range of systemic errors in what must be the most thorough investigation of its kind ever. Pressures on the crew to fly included an unserviceable Auxiliary Power Unit and lack of ground start facilities at Dryden which meant that the aircraft had to be *hot refuelled*. The weather was deteriorating and both passengers and crew were eager to get home in time for Thanksgiving.

Compounding factors included insufficient information to the crew regarding cold-soaking of the wings, poor communication between cabin and technical crew, a lack of ground support and a declared emergency landing by a Cessna 152 which delayed the F-28's take-off. Well before the accident, a decision had been made to use an inferior de-icing fluid in Canada, on the grounds of cost, which had a much shorter holdover time than the fluid used by the Europeans. (Reason, 1992)

Whilst the accident provides an excellent case study, thanks to the thoroughness of Moshansky's investigation, it is not unique. The apparently vast number of errors was not indicative of a particularly bad accident, rather, a thorough investigation. Very simply, the crew assessed all of the risks that they perceived and decided to take off; they did not expect to crash. It is easy to suggest with hindsight that warnings were there, but difficult to criticise individual active failures in isolation. Although two passengers expressed their concern about snow on the wings to the Flight Attendant, she did not have sufficient training to recognise this as a problem or feel that she should communicate it to the Flight Deck.

The Effect of Terrain

The physical geography of a particular area affects aviation in several ways. Firstly in terms of its effect on environmental conditions, secondly, in terms of its effect on collision risk and finally, as will be considered in the next chapter, the effect of spatial separation on routes and aircraft range.

Every airport has a unique set of challenges for a pilot on both take-off and landing. This is in terms of surrounding high terrain, buildings and local weather phenomena such as low level shear or rotary turbulence. En route, the physical environment also affects meteorological conditions and in certain circumstances, the risk of collision. Finally, the geography of the area in which operations occur will influence facets of the aviation system such as aircraft type selection, sector length and demand.

Relief and the Environment

Physical geography has a direct effect upon weather conditions and although any area is prone to extremes, there are several traits of climatic zones, which, by definition, are prevalent. It may also be a way of predicting extreme phenomena such as tornadoes, cyclones, duststorms or jet-streams. The presence or otherwise of relief induced weather or climatic phenomena within Australia may be a significant component of the natural environment in which aviation operates.

Crowder (1995) observes that 'Australia is an old continent. Much of its area is relatively flat and its mountains are small by comparison with those of all the other continents. The effect is the exceptionally arid central region. Global rainfall averages at approximately 1,000 mm, but 50% of Australia receives less than 300 mm per annum and 90% less than 800 mm'. Skies remain almost annually blue, except for the more inhabited coastal strips. This helps to propagate the myth that Australia is typified by blues skies and *good aviation weather.* However, as discussed earlier, weather conditions within Australia do include a number of extremes that are potentially hazardous to aviation.

The higher level of rainfall around the coastal regions, particularly the eastern seaboard and northern coast is not just a function of its close proximity to the sea. If the relationship were that simple, then the coast along the Great Australian Bight and North West of Australia would not be so dry. (Annual median rainfall of 200 mm). The tropical climatic zone and high coastal terrain are largely responsible for the high level of rainfall in the far northern latitudes and the Great Dividing Range for the average to high levels of rainfall along the eastern seaboard. Strong winds such as the South East Trade Winds and Westerlies gather moisture as they sweep across the Pacific, which is then released as precipitation as they reach high ground.

Even areas of only moderately high terrain are enough to dictate zones of high precipitation. Around Sydney, the Blue Mountains are the main reason for higher average rainfall, whereas in Melbourne it is the Dandenongs. Similar effects are also visible near the Darling Range, Perth, the Mount Lofty Ranges, Adelaide and Mount Wellington, Hobart. These areas account for the overwhelming majority of RPT movements and as meteorological conditions are most significant at take-off and landing, there are considerable safety implications.

Heavy rainfall may be associated with a degradation of visibility, contamination of runway pavements and in-flight icing. However, the effect of terrain is not just on the frequency or severity of precipitation. Windshear associated with rotary shear or mountain lee waves can have a significant effect on flight operations and is found in Australia, but as RPT crews operate to a limited range of destinations where hazards are well known, the overall risk to RPT safety is diminished.

Another meteorological phenomenon associated with relief is the formation of temperature inversions and associated degradation of visibility. Sydney is a prime example where easterly winds meet the physical barrier of the Blue Mountains and Great Dividing Range. As such smog is a common occurrence in the Sydney basin which can significantly affect visibility.

The Effect of Relief on Collision Risk

For many years, the greatest source of fatalities in civil aviation has been as a result of controlled flight into terrain (CFIT). A controlled flight into terrain accident occurs when a serviceable aircraft collides with terrain (or water) when still under the apparent control of the flight crew. A survey by Boeing of worldwide airline fatalities between 1988 - 1993 (aircraft over 60,000 lbs take-off weight) revealed 1,883 deaths due to CFIT from a total of 3,513 for all causes (just over 53%), even though such accidents only accounted for 28 of the 76 accidents which were studied (Hughes, 1994).

Reiner (1992) observes that; 'Most CFIT accidents occur in terminal areas, more so on arrival than departures, and often very close to the top - within 100 to 200 feet - of terrain features'. He also goes on to suggest that; 'In every jet transport CFIT accident to date (1990), impact occurred with enough excess energy to have cleared the obstruction if the warning and pull-up had been sufficiently timely.'

The question here is whether the risk of CFIT is any different for the Australian carriers, or within Australia? Is the height or profile of terrain in Australia a factor in the lack of CFIT accidents or are there other, more pertinent factors?

Hughes (1994) suggests that; 'Factors which increase the risk of having a CFIT accident include flying into airports located in mountainous terrain that have only non-precision approaches and no radar coverage.' He adds that in the US and Europe, these destinations tend to be served by secondary level, regional carriers, and as a consequence, the accident rate for this type of operation is two or three times that of large jet operations. The level of terrain, however, is only one factor in the prevention of Controlled Flight Into Terrain (CFIT) accidents. Around 46% of CFIT accidents have occurred where terrain is relatively flat (less than 1,000 ft. higher than airport elevation)(Boeing, 1994) and the most significant factor, according to Boeing, is a lack of Minimum Safe Altitude Warning (MSAW) equipment.

There are few parts of Australia that may be considered to be mountainous. Indeed, it is the flattest continent on earth with only about 2% of terrain over 1,000m above sea level. This is predominantly in the Snowy Mountains area of South West New South Wales, Australian Capital Territory and North-East Victoria with an additional mountainous area in Tasmania. The only major RPT airports that are situated close to high terrain are Canberra, which is surrounded by mountains; and Cairns, which has a large escarpment to the west.

The reporting of nuisance GPWS (Ground Proximity Warning System) alerts on approach to Canberra is very high. (A nuisance warning is where the GPWS is *fooled* into making an alert by rising terrain, especially on approach where the terrain about to drop away before the aircraft would collide. They are valid warnings, but are classified as nuisance as the aircraft is on a stable approach path.) Whilst such warnings do not mean that the aircraft is necessarily at threat, they do cause some concern in terms of complacency. The warnings generally occur because of rising terrain whilst the aircraft is descending at relatively high speed and as the GPWS is a non-intelligent system, it cannot see that the terrain is about to drop.

It is possible to avoid these warnings by restricting speed to 200 knots and requiring gear to be lowered earlier. (Quinn, 1997) The latter measure will prevent any GPWS warnings and the former will reduce the number of warnings caused by excessive approach speed. Whilst this may reduce the *cry-wolf* effect of recurrent nuisance warnings, it also disables a safety system which is in fact working correctly and increases the risk from an unstabilised approach.

Whilst the issue of high terrain is obviously relevant to collision risk at airports such as Cairns and Canberra, that risk is also influenced by other factors. These include the prevailing meteorological conditions - Canberra is prone to low level cloud and fog, and Cairns is prone to tropical storms; reaction to GPWS warnings, especially in areas which experience a high volume of nuisance warnings and cancellation of GPWS functions e.g. by early gear deployment. Slower approach speeds may mean that there is more time to recognise and respond to an unstabilised approach, but also mean that there is less kinetic energy available to clear an obstacle.

Flat terrain CFIT accidents, such as the loss of an Eastern L-1011 on approach to Miami in 1972 when the flightcrew were distracted by a failed warning bulb and did not realise that the autopilot had been disengaged, illustrate that CFIT accidents are not just a function of terrain. In other words, the aircraft was on such a trajectory that it was going to hit the ground regardless of the height of the terrain. The crucial factor in these accidents is crew performance, both in terms of vigilance and situational awareness.

In aircraft where GPWS is fitted and working, crewmembers must take a decision to ignore warnings if an accident is to occur, except in instances of particularly steeply rising terrain. In a multi-crew environment, effective crew resource management (CRM) aims to improve the quality of communications and decision making. 'Provided that all the relevant information was available to them, it is difficult to see what could be done to resolve this particular problem other than emphasise the potential consequences during training and whenever else an opportune moment presents itself.' (CAA, 1982) The goal of effective CRM is to ensure that the right information is communicated to the right person at the right time and in the right way. CRM was not introduced as a CFIT evasion strategy, yet, when used properly, it can have the effect of reducing the number of such occurrences.

In November 1983, a B-747 collided with terrain during an ILS approach to Madrid Airport in spite of a serviceable GPWS. The Spanish Accident Investigation Board found that; 'The pilot in command did not take the required corrective action when the GPWS alarm signals were activated.' (Doss, 1990) and cited as contributory factor the '...failure of the crew to take corrective action in accordance with the operating instructions of the ground proximity warning system.' In other words, the accident could have been avoided through at least two prevention strategies; namely better GPWS training and better CRM training.

The fact that CFIT type accidents have continued in spite of GPWS warnings serves to highlight the fact that a safety countermeasure can be undermined if behaviours are adapted to compensate for it. If the fitment of GPWS means that less attention is paid to stabilising approaches, then the safety strategy has not worked and risk homeostasis has occurred. Similarly, if GPWS warnings are ignored or avoided through procedures such as dropping landing gear early, then the usefulness of the system is degraded.

Collision risk is obviously linked to terrain, but is not necessarily a function of high terrain. Exceptionally flat approaches are able to create an optical illusion, which can lead to undershoot collisions with terrain. This phenomena can also occur at airfields with exceptionally long runways.

Australia may have relatively low terrain and operations do not experience the same level of complex terrain as in, for example, Papua New Guinea or South America, but this does not mean that operators are immune to collision with terrain. Hazards do exist, both within Australia and for Australian carriers operating overseas.

Although the general perception is that Australia is blessed with good aviation weather and relatively flat terrain, this is only part of the story and not the sole reason behind the safety record. The nation's climate includes temperate, subtropical and tropical zones and conditions that are hazardous to aviation, such as microbursts and windshear, are not uncommon. There is certainly no room for complacency: The accident record clearly demonstrates how unforgiving the natural environment can be.

5   Exploring the Human Geography

The Effect of Spatial Separation

Australia is a country of sparse population. Roughly the same size as the United States at 2,966,368 square miles, it averages only five people per square mile. Eighty percent of the population live in the cities. The main six centres of population being Sydney, Melbourne, Brisbane, Perth, Adelaide and Canberra which account for about 10 million of the country's 16 million inhabitants (National Geographic, 1988). In comparison, the population of the USA in 1997 was estimated to be 260 million.

The Effect on Historical Development

Modern day Australia was discovered by Captain James Cook in 1770 and colonised as '...part of Europe - New South Wales, as Cook called it - not of Asia' (Terrill, 1988). It was used as a British penal colony between 1788 and 1868, which saw 162,000 convicts transported from the mother country.

Blainey (1977) tackles the importance of distance to the development of Australia in his book *The Tyranny of Distance*. He prefaces the text with a valuable warning;

> Distance - or its enemy, efficient transport - is not simply an explanation for much that happened in Australia's history. Once the problem of distance is understood it becomes difficult to accept many of the prevailing interpretations of other events in Australia's history. Distance itself may not explain why they happened, but it forces a search for new explanations...
>
> ...It illuminates the reasons why Australia was for long such a masculine society, why it became a more equalitarian society than North America and why it was a relatively peaceful society.

Although the focus of Blainey's text is ostensibly the transportation challenges of the 18th century, it provides an interesting historical insight that provides some clues as to why the aviation industry developed in the way it did.

Modern Australia's Anglo roots proved to be incredibly important in the way it developed since colonisation, not least in crafting the Australian culture. Although many traits represent an imported Anglo culture, there were certain new traits that represented. This was a function of the type of people that were transported or chose to travel there and also the challenges that were presented to them. For example, that of distance, not just on a world scale but within the country. Blainey (1977) illustrates that whilst Australia was 12,000 miles away from Europe, her coastline also represented a similar distance. As the State capitals are a significant distance apart, long distance travel was an inescapable barrier to trade, especially when compared to domestic travel in the United Kingdom.

The first voyage of convicts in 1788 took six months of sailing time to reach Sydney. Between the early 1850's and 1870's, the average passage between London and Australia shrunk from 90 to 45 days, largely the result of fast steamships, trains and the opening of the Suez Canal.

Table 5.1   **Distances between state capitals (kms)**

Capitals

	Adel	Bris	Dar	Hob	Mel	Per	Syd
Adelaide, SA	--	1622	2624	1260	650	2118	1165
Brisbane, QLD	1622	--	2825	1989	1379	3610	748
Darwin, NT	2624	2825	--	3788	3178	2653	3155
Hobart, TAS*	1260	1989	3788	--	610	3320	1040
Melbourne, VIC	650	1379	3178	610	--	2710	706
Perth, WA	2118	3610	2653	3320	2710	--	3283
Sydney, NSW	1165	748	3155	1040	706	3283	--

* Distances calculated via Melbourne

Source: Adapted from Ansett Australia In-flight Magazine.

The Coming of the Railways

Within Australia, transportation developed slowly. Navigable inland waterways were few and far between and prone to periods of prolonged drought. Coastal shipping benefited from the development of the steam ship but this remained a slow method of intra-state travel due to the distances involved. The arrival of the railways in the mid 19th Century brought a new optimism for fast transport, but its true potential was never realised because of the high cost of permanent way. The small population (by 1860, the population was only 1,097,305 (Dyster and Meredith, 1990)) and great intra-state distances meant profitability was poor and development was slow.

By 1881, Australia had nearly 4,000 miles of railway although it was not a linked network. The fourteen different railways companies were all physically separated and operated on a variety of gauges. Even the important route between Melbourne and Sydney stopped at the New South Wales - Victoria border and passengers and freight had to be transhipped. Nevertheless, the 600 mile journey by train reduced the trip to 18 hours which was less than half of the time of a coastal steamer. (Blainey, 1977)

By 1921, the length of railways in Australia was 26,000 miles - in relation to its population, Australia had more railway than any country in the world. However, it was not until 1962 that trains operated all the way between Sydney and Melbourne on a single track gauge and 1995 before the trunk rail route from Brisbane to Perth via Sydney, Melbourne and Adelaide was totally converted to the same standard gauge. Vast distances prevented the railways from attaining the dominance that they achieved in Europe. Even now there are only two passenger trains a day between Melbourne and Sydney, the two largest cities.

The Road Transport Revolution

The first Model-T Fords arrived in 1909 and Australia boasted just 5,000 motor cars by 1910. By 1930, there were nearly 600,000 cars and trucks. Roads ranged in standard from tarmacadam urban and suburban routes to dirt roads, which covered most of the longer distances. Even today, large sections of the road network are dirt tracks. New transport opportunities, low ground rents and an abundance of land led to a suburban sprawl of many cities and towns which is reflected in the number of single storey residential properties compared to the mother country.

Car ownership within Australia became an absolute essential especially in the sparsely populated outback areas. However, the distances involved still placed restrictions on the development of road transport as a long distance transportation mode. Although fuel remained relatively cheap within urban areas, the logistics of distribution meant that operating costs in country areas remained high.

Even now, a car journey between Melbourne and Sydney would typically take in excess of 13 hours. To cross the Nullarbor Plain between Adelaide and Perth is recommended to take a minimum of three days.

The New Opportunities of Aviation

Aviation arrived in Australia in 1910 thanks to the American escapologist, Harry Houdini. Although initially a source of entertainment, the potential of aviation for conquering Australia's great distance handicap was great, even if few realised it early on. Within Australia, the road, waterway and railway networks remained sparsely developed and to reach her trading partners in the Northern Hemisphere, the only method available remained the ship. A second demonstration in 1914 by the Frenchman Maurice Guillaux involved flying a single bag of 2,000 postcards between Sydney and Melbourne in three days. Although a success, the true birth of civil aviation in Australia was not to occur until five years later following the First World War. (Job, 1991)

Wartime brought considerable technological advances in aircraft. Once hostilities ceased, converted military aircraft began operations carrying passengers and mail. In 1919, the Australian Government offered £10,000 to the first crew to fly from England to Australia in less than a month. This challenge was met by Sir Ross and Sir Keith Smith in a Vickers Vimy and raised the profile of civil aviation in Australia. A number of airlines were born following the armistice but it soon became obvious that without some form of subsidy, no one was prepared to invest heavily in such a new and unproven technology.

On 24th November 1920, the Air Navigation Act was passed and the Civil Aviation Branch was created. Lieutenant-Colonel H. H. Brinsmead was appointed Controller of Civil Aviation with his duties being described as; 'The inspection, registration and certification of airmen, aircraft and aerodromes and to advise on matters affecting the organisation of airlines and schemes for the encouragement of civil aviation.' (Job, 1991) The

Government considered several mail routes for subsidy before announcing in 1921 that the route from Geraldton to Derby would be operated, with their backing, by Major Norman Brearley's Western Australian Airways. The airline, better known as *Airways* tendered to operate a fleet of six Bristol Tourer aircraft, which could each carry two passengers and less than 200 lbs of mail. (Edmonds, 1994)

The first airline service commenced on 4th December 1921 with three aircraft. The first accident occurred the next day when one aircraft made a forced landing in rough country, followed immediately after by the second accident (and first fatal Australian airline accident) as one of the other aircraft attempted to establish the ground position of the first. (Buddee, 1978) Brearley immediately suspended services and informed the Controller of Civil Aviation by telegraph;

'Inform you fatal accident. Aviators Fawcett and Broad killed through machine crashing 100 miles north of Geraldton due to Commonwealth not having landing and emergency landing grounds, prepared as agreed. Other planes returned. Brearley has decided not to proceed until suitable landing grounds ready. Geraldton ground dangerous and condemned by Public Works Department.'(Job, 1991)

Although the cause of the accident was not the integrity of the airfield (the Civil Aviation Branch's finding was 'Error of judgment; pilot banked too steeply in landing in rough country'), it was felt that if the first aircraft had been able to make a safe landing, the second would not have executed its fatal manoeuvre in attempting to follow it. Subsequent surveys by Brinsmead's team allowed Brearley to recommence operations in May 1922. Then followed an accident free year, almost complete adherence to timetables and a small profit. For many, this is seen as the true birth of civil aviation in Australia.

Meanwhile another small aviation company was founded in Queensland in 1920 with the intention of competing in the *Great Australian Air Race*. However, when their backer died, the two pilots, Wilmot Hudson Fysh and Paul McGinnes were unable to take part. Instead they were commission by the race organisers to survey the route across Australia covering 2,166 km in about 50 days. (Buddee, 1978) It was this experience which gave Fysh and McGinnes a realisation of the communication problems and potential for aviation in the outback. This was also the point that they met Fergus McMasters who was destined to become their business partner in their airline. Originally formed on 19th August 1920 the airline became

Queensland and Northern Territory Aerial Services Ltd. or Qantas as it is better known. Now over 80 years old, Qantas is one of the oldest established airlines in the world.

It is not the objective of this book to present a detailed history of how the Australian aviation industry developed. However, there are a number of events that have occurred up to the *jet-age* which have a major significance in terms of safety.

The Southern Cloud Mystery   On 21st March 1931, Australian National Airways Avro Ten *Southern Cloud* was lost in poor weather en route between Sydney and Melbourne. The aircraft did not have any form of radio and no-one witnessed it crash. There was no official search and rescue organisation in Australia at that time, so the Deputy Controller of Civil Aviation organised a search which involved 30 civil and military aircraft over 18 days. Despite these efforts, the wreckage could not be found and it was concluded that no occupants could have survived.

The remains of the aircraft were finally discovered in October 1958 in the Snowy Mountains area of New South Wales. Although an official investigation was not conducted at this late stage, it was obvious to the Investigator in Charge that the aircraft had collided with terrain at high speed.

The significance of this accident is two-fold; Firstly, although the accident occurred in the more populous South Eastern area of Australia, the wreckage was not discovered for 27 years. Early in Australia's aviation history, the possibility of survivors being found and rescued was remote. Secondly, the Inquiry held in 1931 in the absence of the wreckage made a series of recommendations, which were to influence the development of civil aviation in Australia. The most significant recommendations being as follows;

- That ... the carrying of two-way wireless ... be made compulsory in aircraft engaged in regular scheduled passenger services. Action should be taken to give immediate effect to this on the Sydney / Melbourne / Launceston service.
- That the Departmental scheme for a ground wireless direction finding organisation be proceeded with and expedited as an urgent measure.
- That endeavours be made to have an additional synoptic chart drawn by the Weather Bureau ... that arrangements be made for observations of current weather at selected points along air routes at 7am daily, and ... the corrected

aviation forecast, together with a statement of actual conditions over the routes, be issued for rapid distribution to all civil aviation terminal aerodromes and RAAF Stations.

Stinson-A Trimotor Brisbane Another significant accident occurred on 19th February 1937 in the MacPherson Range of Mountains in Southern Queensland. A Stinson-A Trimotor operated by Airlines of Australia was en route from Archerfield to Sydney via Lismore with two crew and five passengers when it collided with terrain. Although searches were commenced by air, water and land, they were called off five days later with 'no hope of finding the aircraft'. However, Bernard O'Reilly, a bushman from within the MacPherson Ranges believed that he knew where the aircraft had crashed, based on the last sightings of the aircraft by his neighbours. A week after the accident, he set off alone into the bush to search for the aircraft. Two days later he discovered the wreckage of the aircraft and two survivors.

The story has passed into Australian folklore as a story of courage and determination in the tradition of the Australian bush and yet its significance as a cultural indicator is not to be underestimated. The secondary implication of the accident was the reaction of the Australian public to what was almost another *Southern Cloud*. Although, recommended after the loss of the *Southern Cloud*, civil aviation requirements still did not require the fitment of radios on airliners. Only three days after the *Brisbane* disappeared, the Minister of Defence announced that radio beacons would be installed at the principal airports of Australia.

Douglas DC-2 Kyeema In 1937, Australian National Airways (ANA) acquired controlling interests of Airlines of Australia. Operating under the latter company badge, DC-2 *Kyeema* was scheduled to operate with an ANA crew between Melbourne, Adelaide, Sydney and Brisbane on 25th October 1938. On approach to Essendon airport from Adelaide, the aircraft collided with the slope of Mount Dandenong with the loss of 18 souls. Although the aircraft was fitted with a primitive radio, it was not yet able to use the Lourenz radio range finding system, which was still being tested nearly two years after the *Brisbane* accident. The investigation focussed on the role of the radio contact between aircraft and the ground and suggested that the reporting system did not adequately prevent accidents. (Official Accident Inquiry cited in Charlwood, 1981)

At an early stage in this inquiry, it appears to us that a serious element of risk existed whenever miscalculations or errors of observation on the part of the pilot occurred during a flight, and it was suggested that it would be a simple matter to introduce a system whereby the movements of aircraft could be checked by a competent person on the ground.

It was following this accident that the Air Traffic Control system was created in Australia. Radio Ranges, aeradio stations and direction-finding units were introduced and supplemented by a uniquely Australian feature, the Flight Checking Officer.

Douglas DC-3 Lutana   By 1948, civil aviation had undergone significant technological advancement as a result of the Second World War and the Australian ATC system had benefited from an ICAO conference held in Melbourne in 1947. However on 2nd September 1948, ANA DC-3 *Lutana* struck a mountain near Nundle in New South Wales en route from Brisbane to Sydney with the loss of 13 souls. The aircraft was under the control of radio ranges at Brisbane, Kempsey and Sydney and flying at night. After overflying the town of Kyogle, the aircraft drifted west of its track, but this fact was not detected either by the crew or ATC. When the aircraft was given permission to descend to 4,000 feet on its way into Sydney it was in fact 100 miles off track in mountainous terrain.

The subsequent inquiry criticised the ATC system for paying too much attention to aircraft separation rather than keeping the aircraft clear of terrain. Charlwood (1981) cites the official inquiry chairman; 'It appears to me that a criticism that can justly be levelled against the Department is that it is not using modern scientific aids which were discovered and used during the war, that they are requiring information from the pilot which should be known to them or which could be obtained by them accurately and instantaneously from the ground without choking the communication channels. I have in mind such scientific aids as radar, and automatic position finding by cathode ray direction finders.'

Fokker F27 Abel Tasman   Although a number of accidents occurred following the loss of the *Lutana*, the next highly significant accident involved a TAA Fokker F27 on 10th June 1960. The aircraft was scheduled to fly from Brisbane to Mackay with stops at Maryborough and Rockhampton and departed the latter with four crew and twenty-five passengers, which included nine schoolboys on their way home for the Queen's Birthday long week-end. On approach to Mackay, weather conditions deteriorated and the aircraft held for forty minutes at 5,000 feet awaiting an improvement in conditions. (Job, 1992) When eventually cleared to land with fog lifting and ground visibility up to three miles, the aircraft struck the sea with the loss of all on board. Although no definite cause for the accident was ever reached, the significance of the accident is that it remains the worst accident involving an Australian aircraft in terms of lives lost. Further, the tragic loss of nine schoolmates shocked the nation and has never been forgotten.

Trading Partners

Australia's links with the mother country, England, were incredibly strong during the late 18th and 19th centuries. This was not just in terms of the incumbent population and culture but also in terms of the import and export of products. Trade figures for the late 19th century show just how much Australia was forced to rely on the UK (Dyster and Meredith, 1990) and other far distance trading partners. In terms of imports, 76% of trade was with European countries and only 7% were specified as Asia or New Zealand.

By the early 20th century, although the UK's dominance as the premier trading partner had faded, the majority of trade remained with the North Atlantic countries. When Qantas was founded in 1920, approximately three quarters of the international trade remained with the USA, Canada and the states that later formed the EC. Unsurprisingly, this raised the importance of aviation as a transport mode. 'Australia, with a small population spread over a very large island and geographically isolated from most of the rest of the world, has always relied heavily on aviation both domestically and internationally' (Felton, 1988).

The trade potential for aviation was great and was recognised initially for *Air Mail* services. Indeed, it was funding for the delivery of mail that allowed the first operators to develop, not least because early aircraft had a very low capacity. For example, the Armstrong Whitworth FK-8 used by Qantas from 1922 carried just three passengers.

The international route between London and Australia opened in 1934, operated by Qantas and Imperial Airways. 'From such an endeavour was borne something of enormous importance to this remote isolated continent.' (Buddee, 1978) Freight was primarily mail, but the vastly reduced travel time between Australia and England meant that trade visits could be more easily made. Within Australia, the Air Mail and passenger business grew and was supplemented by some unusual uses of aircraft. For example, an operation called *Air Beef* was set up in 1949 to transport meat from an inland slaughterhouse to the coast at Derby for shipment. This service was developed because local roads were so poor and it lasted until 1961.

However, major threats to the success of these early services were reliability and economics. Yields were very low and operations generally required Government subsidy to survive. The loss of aircraft represented a significant cost, not least because replacement airframes were imported from the US and UK by ship. Aircraft reliability was critical and was compounded by the spatial separation of Australia.

If an aircraft became lost on a flight, the chances of rescue were limited. If the aircraft was flying close to telegraph wires, the pilot carried a primitive transmitter, which could be connected to the wires to call for help. If the aircraft landed away from telegraph wires then the sparseness of the Australian population often meant they could not be located. Poor engine reliability, weather conditions and navigation often dictated that early aircraft would make landings away from airfields. This was less of a problem in Europe and North America where the population density was higher and there was a greater chance of help being close by. Australia was in the unenviable position of being a vast country with meagre population density and surrounded by an even bigger ocean.

This meant that an above average level of reliability was needed for Australian operations to maintain the efficiency and safety of its operators compared with those in Europe and North America. Demands for high reliability were particularly vociferous from those within the industry and led to the development of an industry culture that still exists in one form or another today. The potential for trade was great and the potential for catastrophic failure was higher. Whilst this may have been expected to lead to a culture of corner cutting in order to deliver results, the role of key individuals in the development of airlines such as Qantas and Ansett assured that the opposite was true. The strong influence of dedicated aviators and mechanics with a desire to do things *right or not at all* set a standard that has become an expectation.

In recent years, aviation has become almost the sole form of international travel to and from Australia and therefore has reached a political and social position that is still even more critical than many developed countries around the world. Fast high-speed train and ferry links have maintained a competitive edge with aviation in many countries by virtue of airport location, congestion and the cost of air travel. This is not the case for Australia, which sustains an expectation of efficient and safe air operations. In turn this must translate into a deliverable on the part of the airline if they are to avoid a significant political backlash.

Route Structure

Two potential problems highlighted by Barnett, Abraham and Schimmel (1979) in comparing safety records, were that different airlines fly routes of different lengths and different airlines fly into different airports through different airspace.

The authors believed there to be no reason to give special consideration to flights of different length relative to the risk of an accident. Their conclusion was in response to a statistical exercise to look at 46 accidents involving US Domestic airlines between 1957-76.

Boeing's accident summary (1996) plots a profile of hull loss accidents against stage of flight. The data shows that although the cruise phase is by far the longest component of any flight, the majority of accidents occur within the shortest phases. In the average 6% of a flight that is takeoff, initial climb, final approach and landing, a staggering 68.6% of accidents are recorded. By this token, aircraft types that are used for short haul may appear to have significantly worse accident rates than their widebody counterparts. This is especially so if accident rates are compared in terms of flying hours or simply annual accident rates. Caesar (1994) suggests that the risk exposure for a short-haul B737 flying 10-15 sectors a day is approximately four times higher than that of a long-haul B747 flying 3-4 sectors.

This is a function of several factors including;

- Airframe stress of recurrent pressure cycles
- Proximity of hazards on takeoff, climb, approach and landing
- Aircraft operating at extremes of performance envelope on takeoff and landing
- Greater percentage of time spent on high workload tasks.

Short sectors work the aircraft harder. The effects of pressure change as aircraft ascend and descend place stresses on the integrity of the airframe. Operating at lower altitudes also exposes aircraft to greater corrosion, especially in the saline air associated with *island hopping* flights.

In April 1988, an Aloha Airlines B737 lost a large section of its roof on a flight over Hawaii. The airframe had completed 90,000 cycles (the second highest by any B737) and had been used exclusively for Hawaiian island hopping throughout its life where flights were, on average, 20 minutes long. The repeated pressurisation, depressurisation and exposure to corrosive sea water had led the aircraft to disintegrate in flight. Against the odds, the aircraft survived to land in spite of severe structural damage.

Crews are also exposed to a far greater risk of performance overload in the critical takeoff and landing phases. 'Crew workload is one of the most important human factors issues in aviation. Sustained high workload levels will overtax the crew, thus decreasing the margin of safety and increasing the probability of an accident.' (Blomberg, Schwartz, Speyer and Fouillot, 1988, Speyer and Blomberg, 1989)

However, it cannot be inferred that mistakes are made proportionally to workload as this is not the case. '...It is errors that cause aviation accidents, not workload' (Blomberg et al., 1988) Errors can be made at both extremes of workload - underload and overload and anywhere in between. 'The optimisation rather than the reduction of flight crew workload is recognised as a major goal of human factors engineering.' (Braby, Muir and Harris, 1991) Short sector length will expose air crews to more of the high workload phases of a flight than long haul which may reduce the danger of task overload, but conversely, prolonged periods of relative inactivity (i.e. cruise) followed by a period of high workload such as during approach and landing may also cause alertness problems. Does this benefit crews in providing experience of these important areas of flying or will crews become blazé to the hazards through complacency?

Caesar (1994) suggests it to be the former; '... (short-haul) crews have a better arousal level, there is less chance for boredom, they are continually engaged in meaningful work. Their aircraft are more compact, less critical to handle with two engines overpowered by definition. The crews stay constantly highly experienced in the manual handling as well as the automatic flight guidance system programming due to more frequent take offs and landings.'

Barnett, Abraham and Schimmel's study (1979) questioned whether flight-length was a significant factor in the accident record. For each accident the length of sector was recorded to provide a flight-length distribution which could then be compared with the actual distribution from the 1975 Official Airline Guide. In testing their hypothesis, the authors concluded that '...collectively, the US domestic airlines perform their longer flights at about the same level of safety as their shorter ones.' This, they felt, provided no basis for special consideration to airlines with unusually long routes. Unfortunately, the conclusions of the MIT group are perhaps not as robust as they may initially seem.

The data used (1957-76) cover a period of rapid change within the industry. During this time, the accident rate varied considerably as airlines introduced jet aircraft. This change included the early Comet metal fatigue disasters and a number of transition accidents where crews did not successfully make the change in style from flying propeller driven aircraft to jets. The industry was also still playing host to a number of ex-WWII pilots and the *gung-ho* or *wrong stuff* habits that they often brought with them.

More importantly, the period coincided with the introduction of new aircraft to different lengths of route and the cascade effect this had on operations. For example, the B707, which was introduced in the late 1950's as a front line, long haul jet, was being replaced by widebody types such as the B747, DC10 and L-1011 by the early 1970's. The B707's would be cascaded down to shorter haul flights that may have been operated by older aircraft. By this time, the aircraft were known technology and not prone to some of the mishaps or failures that were experienced during the earlier, longer haul, years.

The sample of accidents used also fails to explore the rather wider issue of international operations. The USA is a closed system in terms of a unitary regulator (the FAA) and single language. This would mean that variations in routes, ATC quality and aerodrome facilities are all within the minima set by the FAA, or in the former case, by geography. Admittedly, the quality of facilities is not a function of the length of a route, but longer routes do tend to include the crossing of state boundaries and the differences in regulation that this inevitably means. It is difficult to represent the relationship between distance and risk exposure for this reason. A journey of 1,000 miles may be very safe or very dangerous depending on where the aircraft has to fly over or to.

The extremes of short-haul and long-haul are not fully represented by the American domestic market. The longest coast to coast flights do not push aircraft to the same performance limitations that, for example, a 13hr 40min Singapore-London sector does. On such flights, the danger comes not just from the performance of the crew or the extended reliability required from the aircraft, but also from the fact that the aircraft will be taking off with a full fuel load and on many routes, a full complement of passengers and freight. Elements of the flight that cross large expanses of water or over *third world airspace* also increase the level of hazard on long haul flights in a way that is not experienced by the US domestic market.

Time zone differences are also important. The difference between Pacific Time (GMT-8) and Eastern Time (GMT-5) is only three hours and has a limited effect on circadian rhythms. However, a longer sector such as London to Singapore (+8 hours) or a transmeridian flight such as Auckland - Los Angeles will have a profound affect on sleep patterns and performance levels. Graeber, Dement, Nicholson, Sasaki and Wegmann (1986) and Nicholson, Pascoe, Spencer, Stone and Green (1986) noted the effects of jetlag on performance and physiological state, which included disruption to the sleep-wake cycle, insomnia, tiredness, gastric problems and longer term sleep deprivation. (Cabon, Mollard, Coblentz, Fouillot and Speyer, 1991). Stone's (1991) examination of London - Tokyo flights suggests that on return from such long haul trips, crews can take anything up to six days to resynchronise their body clocks and get back to normal performance.

Heavy crewing is one strategy that is employed to offset the effects of tiredness and fatigue on long-haul flights. This is a procedure whereby an extra First or Second Officer is carried to act as a *Cruise Pilot* and allow the other pilots to take a rest break. Flight and Duty time limitations imposed by the regulator make such a procedure necessary although it should be noted that Qantas exceed this requirement by 80% (Qantas, 1997). Custom-built rest areas on long-haul aircraft such as the B747-400 allow crewmembers to sleep and freshen-up in flight to maintain vigilance and performance. Graeber, Rosekind, Connell and Dinges (1990) conducted a study of controlled napping which '...clearly demonstrated that a pre-planned cockpit nap was associated with significantly better behavioural performance and higher levels of alertness.' In recent times, Qantas has adopted an official policy of controlled napping and as Graeber et al. (1990) note, 'there may be considerable safety benefits from such a sanctioned policy.'

Caesar's work, (1994) mainly with Lufthansa, where he was their first safety pilot, goes further to state that 'Since the risk is not the time in the air but the frequency of take-offs and landings, the Australian airlines including Qantas run a lower risk.' However, this seems contrary to Caesar's view that long haul flights have an unfavourable record because of the extreme arousal-alertness concept where prolonged periods of boredom are followed by burst of intense activity. To this end, he notes the way that airlines such as Qantas train to account for a lack of experience at '...less routine handling' of large aircraft.

Qantas pilots operating the B747-400 on routes such as Sydney-Los Angles, Frankfurt or London may only make one or two physical landings in a month due to the number of crew operating the aircraft and the large number of flying hours accumulated in single sectors. The lack of hands-on experience is supposed to be mitigated through experience in the flight simulator. CASA requirements for simulator training equate to one check per year (in addition to one route check) whereas Qantas policy is for flight crew to complete three simulator checks per annum. This exceedance costs the airline approximately $12 million per year. (Qantas, 1997)

International crews are therefore forced to fly in a variety of conditions with limited hands-on experience. For a given pilot, this may mean that of the two landings he makes in a month, one may be in a hot, tropical storm in Singapore or Hong Kong and the other may be in foggy and icy conditions at Heathrow or Frankfurt. The suggestion is that the long-haul nature of the flying only heightens the criticality of take-offs and landings which negates the *positive averaging* effect of flying long cruise sectors.

Australian carriers fly a wide variety of routes using jet aircraft, which cover both short sectors (30-40 minutes) and ultra long sectors (13-14 hours). It is all but impossible to equate sector length to risk as some of the risk of short sectors (e.g. excessive workload) are balanced out on long sectors (e.g. circadian disrythmia). However, various risks associated with e.g ultra-long distance flight have been ameliorated through safety strategies such as extra training and heavy crewing. Therefore, although active failures caused by deficient vigilance or excessive workload may not be easily proven to be minimised in the Australian system, higher-level safety strategies may reduce latent *decision-making* failures. For example, policy decisions regarding the exceedance of regulations point to an aspect of high-level safety culture over and above commerciality.

A second point considered by Barnett, Abraham and Schimmel's study (1979) was in regard to the *fixed infrastructure* of the aviation system;

namely airports and air traffic control areas. In other words, 'different air-lines fly into different airports through different airspace'.

Every airport and piece of airspace brings with it a unique set of problems for any aircraft. These may be in terms of approach aids, runway length, local weather conditions and approach path or in terms of traffic density, control procedures, uncontrolled traffic and quality of air traffic control.

Although there are minimum requirements set by international agreement to be met before aircraft can be operated into particular areas, there is usually a *margin of safety* between requirement and provision. For example, in terms of runway length, an aircraft would not generally be operated into an airfield where the declared distances were exactly the same as its required minima. It is more likely that there will be some extra margin, which provides a safety net above that required by law.

Airlines have a choice of the routes they fly and airports they use. If an airport is deemed unsafe then it is the airline's prerogative not to fly there. Barnett et al. (1979) consider it '...unreasonable to treat airlines as the hapless victims of airport conditions beyond their control' and count accidents at *hazardous areas* in their analysis of airline safety records.

The International Federation of Airline Pilot Associations (IFAL-PA) has an *Annex 19* of airports awarded *black stars* for dangerous deficiencies. (Barlay, 1991) The list covers airports all around the world but is strictly confidential due to its politically sensitive nature. However, it is known that neither Qantas nor Ansett fly to any of the airports mentioned on the list (Lewis, 1995).

The important question to answer is whether airlines make decisions about whether it is safe to fly into a particular airport or whether airport selection is solely the domain of the commercial department. The Vice President (Operations) of Britannia Airways (Grieve, 1994) explained that in their case, the decision about where to fly was made mainly by market forces. The message was that if they did not operate to a particular destination, then another airline would. An example of an airport where crews would prefer not to fly is the 'stone aircraft carrier' at Funchal, Madeira. The runway starts and ends with a sheer cliff, which allows little room for error. From a purely operational point of view, the airlines would probably choose not to operate large jet transports to this destination, but from a commercial point of view, this is a key location for the European holiday market and airlines such as Britannia have little choice but to operate there.

Qantas' international routes represent premier destinations, which are often the country's capital cities. As such, the level of service is usually already high and there has been little need to exercise any sort of safety-led right not to fly there. In an incident where a strike at Singapore Changi airport meant that the RFFS cover was below ICAO minima, Qantas did suspend operations when other airlines did not, but this was a relatively isolated incident. It may, however give an important indication of the difference in safety attitude between Qantas and other international carriers at the time.

The tyranny of distance has provided many challenges for Australia as the country has developed. The consequences have been both positive and negative. The vast distances and low population density raised the stakes of failure for early aviators but, this seems to have created a culture of expected reliability. The reaction to a number of high profile accidents and the hazards of long distance operations has built the system that appears to have worked so well through the jet age. Central to this development has been a number of key personalities and from them, an industry culture which continues to exist today. Issues of culture will be explored in the next chapter.

6 Culture is not Spelt with a "K"

Introducing the Concept of Culture

Hofstede (1980) defines culture as 'the collective programming of the mind which distinguishes the members of one state from another' and then proceeds to say, 'culture is to a human collectivity what personality is to an individual.'

Trompenaars (1993) refers to several different levels of culture, the highest being that of a national or regional society. Within particular organisations, attitudes may be described as a corporate or organisational culture and within specific functions, a professional culture. Although the focus of his work is primarily oriented towards national culture, consideration of Australian aviation safety covers areas of national, organisational and professional culture. It is the contention of this case study that there are facets of the aviation industry culture that span and influence a number of apparently different corporate cultures. The expectation is that not only will there be certain cultural traits specific to the aviation industry at world and national levels but also to specific roles within that profession such as pilots, air traffic controllers etc.

As Figure 6.1 demonstrates, although a national culture can influence an entire industry culture and in turn the industry culture may influence both organisational and professional cultures, there are multiple cultures that can exist within any one group. That is to say that there will be several organisational cultures within any industry culture and there may also be several professional cultures.

As an example, Figure 6.1 may be applied to the Australian aviation system. The organisational cultures may represent those of Ansett Australia and Qantas Airways and the professional cultures may represent flight crew and maintenance crews. Just as maintenance and flight crew of, say, Ansett Australia will share certain organisational culture traits, so too maintenance crew of Ansett and Qantas will share other professional traits.

Figure 6.1   The multiple layers of culture

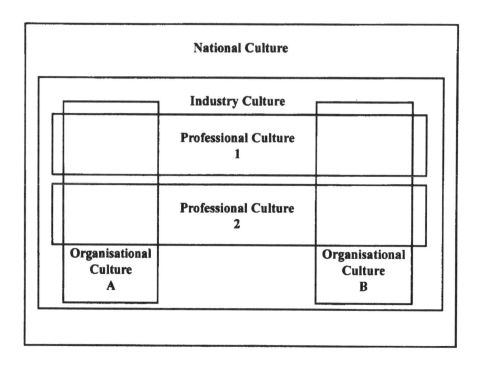

Merritt (1993) simply defines culture as '...the values and practices we share with others that help us define us as a group, especially in relation to other groups.' Tylor (1924) suggests that culture may be defined as 'that complex whole which includes knowledge, belief, art, morals, law, custom and any other capabilities and habits acquired by man as a member of society'. Essentially, culture is a form of subtle mental programming that may only be apparent when compared to other, different, cultures.

For many however, culture seems very visual and easily defined characteristic that is represented through the artistic disciplines of music, visual arts and literature. This perhaps develops a misconception that certain groups can have "more culture" or even "no culture", whereas the reality is that all groups have a distinct and unique culture that may be represented in different ways. These areas of "popular culture" do not tell the full story of the multiple elements that make up any one cultural group.

Culture becomes part of a child's education from as soon as they are capable of learning and is highly influential in the way it develops and becomes educated. It helps define both their attitudes to, and expectations of, other people and interacts with their personality to develop links with other groups (and therefore compatible cultures).

The importance of culture and communication as influencing factors is highlighted by Gudykunst and Ting-Toomey (1988) who write; 'Communication and culture reciprocally influence each other. The culture from which individuals come affects the way they communicate, and the way individuals communicate can change the culture they share.'

Hofstede and Trompenaars' Indices of Culture

Trompenaars (1993) suggests that the exploration of culture should distinguish differences on three levels; our relationships with other people; the passage of time and the influence of the environment. Using a series of questionnaires, the author surveyed 30 companies in 50 countries to compile a database of 15,000 participants' answers. The sample represented 75% from management positions and 25% from general administrative tasks.

Although Trompenaars had spent some time working with Hofstede, he clearly admits that they do not always agree on the nature of culture. What is clear however, is that there is more agreement than disagreement in their coverage of the subject and it is of great value to the reader to explore their work further before concentrating specifically on aviation.

In his work, Hofstede conducted probably the most widely cited study of cross-cultural psychology, studying 117,000 computing industry employees in 66 countries through the HERMES program. Hofstede suggested four areas of cultural difference that can categorise national cultures. For each of these areas, he developed a scale by which to rate each country on a scale between 1 and 100. The indices are described as follows:

Individualism (IDV)   This is the 'relationship between the individual and the collectivity that prevails in a given society'. While different species are seen to be gregarious (e.g. wolves) and others solitary (e.g. tigers), Hofstede suggests that even within a species, there are variations. Although humans may be classified as "gregarious" there are differences in, for example, the complexity of family units. This extends into human society through societal norms in the form of value systems of major groups of the population. These institutions may be educational, religious, political and utilitarian, for example.

Within the context of a particular organisation, individualism may be revealed through the relationship between a person and that organisation. For example, a more collective culture would call for a "greater emotional dependence of members on their organisations". Johnston (1993b) simplifies the dimension by describing strongly individualistic cultures such as the USA as giving primacy to '...personal initiative and individual achievement' and strongly collectivist cultures as displaying '...much tighter and social obligations to clan, class or group.'

The results index to describe the level of individual within each country is based on work goals survey data and is represented as a simple mathematical transformation of the results onto a scale between zero and 100. The highest individualism values are found in USA, Australia and Great Britain; the lowest (and therefore most collectivist) are found in Panama, Ecuador and Guatemala.

Power-Distance (PDI)   This index basically concerns itself with human inequality. In society, this can occur in a number of different ways from social status to wealth and from power to laws and rights. The index used by Hofstede to illustrate this inequality is called power-distance and is 'a measure of the interpersonal power or influence between boss and subordinate as perceived by the least powerful of the two; the subordinate.' Hofstede's work in this area is in fact derived from that of Mulder (1980) who defines power-distance as 'the degree of inequality in power between a less powerful individual (I) and a more powerful other (O), in which I and O belong to the same (loosely or tightly knit) social system.' The suggestion is that power distance within an organisation is, to a considerable extent, determined by national culture.

Hofstede's data show countries such as Malaysia, Philippines and Mexico exhibit high "Power-Distance" whereas countries such as Australia, Great Britain, Ireland and New Zealand demonstrate a low effect. Johnston (1993b) adds that in countries with a high power-distance '...social inequality is readily accepted.' Leaders are expected to be decisive and subordinates are expected to know their place.

Uncertainty Avoidance (UAI) This is another trait of national culture, which describes the variability of tolerance for uncertainty within a nation, based on three indicators; rule orientation, employment stability and stress. All individuals and organisations face levels of uncertainty, which are a function of the environment they live and work within. Different cultures have different values of the importance of avoiding uncertainty and different ways of accomplishing this avoidance. Hofstede's data places countries such as Greece, Portugal and Japan as displaying high level of uncertainty avoidance, with countries such as Sweden, Denmark and Singapore Australia is ranked within the third quartile range and away from USA, Canada and Great Britain, which are all countries it has been close to in the individualism and power-distance.

Countries with a high uncertainty-avoidance score tend to seek clarity and order in society to a level where it becomes intolerant or inflexible. On the other hand, countries with a low UAI score tend to be more adaptable and tolerant by accepting uncertainty as a fact of life. Hofstede suggested a high correlation between PDI and individualism and Smith, Dugan and Trompenaars (1994) speculate that '...empirically speaking, they may be manifestations of the same underlying dimension.'

Masculinity (MAS) Fundamentally, the cultural issue here is 'whether the biological differences between sexes should or should not have implications for their roles in social activities' (Hofstede, 1991). The pattern is based on a "masculine" tendency to be assertive and a "feminine" one to be nurturing. It is not a statement on sexual politics as it is concerned with the "personality" of a culture defined within those two polar descriptions of masculinity and femininity. Highly masculine countries include Japan, Austria and Italy with Australia ranked in the third quartile. Low masculinity or "feminine" cultures include Netherlands, Norway and Sweden.

Johnston (1993b) writes that 'in masculine cultures (e.g. Italy, Australia) ambition and performance are highly valued, and are measured by material success. Forceful behaviour and drive are readily accepted.' This is in contrast to more feminist cultures where there is a stronger emphasis on '...values such as warm social relationships, quality of life and care of the weak.' (Smith, Dugan and Trompenaars, 1994)

The next stage of Hofstede's work was to attempt to integrate the four dimensions. To do this, he carried out a hierarchical cluster analysis which classified the countries into clusters of similar levels of variables.

The result was eight groups formed of 12 clusters. Australia falls within the cluster described as "Anglo" which is described in Table 6.1. This group is probably quite unsurprising and contrasts with extremes of culture, which include two of the Asian groups also highlighted in Table 6.1.

Table 6.1   Sample culture groups from Hofstede (1991)

Anglo	
	Australia
low to medium PDI	Canada
low to medium UAI	Great Britain
high IDV	Ireland
high MAS	New Zealand
	USA

Less Developed Asian	
	Pakistan
high PDI	Taiwan
low to medium UAI	Thailand
low IDV	Hong Kong
medium MAS	India
	Philippines
	Singapore

More Developed Asian	
medium PDI	
high UAI	Japan
medium IDV	
high MAS	

Source: Adapted from Hofstede, 1991.

Hofstede's work, along with the subsequent investigations by Trompenaars have received much critical acclaim in what is a relatively new area of social science. However, the former's work has also been criticised for his sampling frame, which although it covered 53 countries, was concentrated on a single multinational firm, namely IBM. As Smith *et al.* (1994) point out 'Hofstede regarded his matching strategy as a strength, though doubts can be raised as to whether those employed in servicing and marketing in an industrialised nation are necessarily equivalent to those in similar roles in a third world nation'. This is a valid point for discussion that would also translate to the aviation industry: the socioeconomic profile of pilots can be very different. The level of skill and training required to successfully operate a particular airliner may be the same whether the pilot is from a developing or industrialised nation, yet the average earning or GDP per capita may be vastly different.

However, this argument may not be so significant in terms of examining differences between similar occupations in different countries. Hofstede's sample of IBM employees may indeed represent a particular section of society which may not give a completely accurate picture of the full socio-economic range, but at least the data are stabilised by a common profession. The University of Texas Aerospace Crew Research Project's (Merritt, 1993, 1997; Helmreich and Wilhelm, 1997) attempts to replicate Hofstede's work in the aviation field support the validity of his study and to examine its application to the aviation industry. 'It was to test this assumption of universal versus culture-specific pilot behaviour that I undertook the replication study' (Merritt, 1997). The Individualism and Power-Distance indices were selected as being the most relevant to aviation. The conclusion was that '...pilots had higher yet more convergent scores on Individualism than observed in Hofstede's data, a result attributed to modernisation, economic independence, and self-selection into an individualistic profession. The utility of the Masculinity index for the pilot profession was called into doubt. (which is not to question Hofstede's results, but to affirm the unique attributes of the pilot profession)'

Research by others in aviation such as Johnston (1992, 1993), Redding and Ogilvie (1994) and Ooi (Flight International, 1994) has all used Hofstede as a base and hence its inclusion here as a vital starting point for understanding differences. Whilst an entire book could be devoted to the subject of cultural difference and still only barely scratch the surface, the basic introduction of the subject included here sets the scene for the specific examination of the Australian aviation system. Readers interested in a deeper coverage of aviation and culture are directed to *Culture at Work in Aviation and Medicine* by Helmreich and Merritt (Ashgate, 1998).

Culture and Aviation

Aviation is a truly international business. It is difficult to imagine any air operation that does not involve some form of multinational collaboration, whether it is in the design, construction or maintenance of aircraft or the supply of flight planning information, air traffic services or operational crew. As such, the industry is exposed to the full range of cultures on a daily basis, which ultimately affects the way it functions.

Culture is a subject on which many people are self-appointed experts. Experience of other cultures, particularly national cultures, contributes to a level of understanding that is all too often incomplete and laced with misinterpretation. The stereotype is one such unfortunate outcome of cultural exposure, but one that everyone is guilty of using and believing. The mental image of Australians as Crocodile Dundee figures is as real in the minds of Englishmen as the image of pinstriped suits, umbrella and bowler hat in the minds of their antipodean counterparts.

The realms of stereotyping and self-appointed experts are no more apparent than in the aviation industry where exposure to other nationalities is perhaps more common than other professions. It is not fair to say that all stereotypes are wrong, or that expertise based on experience is deficient for the very reason that exposure to other cultures is so high. The aviation industry boasts some great experience in the areas of culture although sheer logistics prevents real multicultural expertise. There is enough anecdotal evidence around the world to tie up a researcher for the rest of their working life if it were to be collected. Even if such a task were to be attempted, the danger will always be that the stories would be collated from a particular cultural viewpoint i.e. that of the researcher.

In April 1996, the IATA Director-General, Pierre Jeanniot addressed the Asia / Pacific Economic Co-operation Working Group aviation seminar in Vancouver, Canada. (Flight International, 1996) Acknowledging the '...cultural characteristics involved in human factors', Jeanniot suggested that, '...it is important that this area be explored professionally in an atmosphere devoid of any emotive cultural sensitivities so that compensating measures can be developed.' It is these "emotive cultural sensitivities" that provide the greatest challenge for researchers both from the point of view of the *researcher* as well as the *researched*. The editorial in Flight International (1996) notes that a study of culture will not be easy. 'It must be international in content, outlook and subject, and it must not be a study by Westerners of the anecdotal cultural differences between themselves and others'.

Early attempts to examine cultural differences, particularly in accident records have suffered somewhat from a lack of data. The fear of offending member states appears to be one of the reasons that international organisations such as ICAO and IATA have been reluctant to publish safety figures that separate member states. As was mentioned earlier in the review of safety statistics, the use of geographic regions is as specific as either seem to go. When a search of the ICAO ADREP database was requested to split accidents / incidents by state of occurrence, the Australian Representative to the Council (Weber, 1994) explained that although the Secretariat would provide what assistance they could, the information contained on ADREP '...is provided by States on the understanding that it is used for accident prevention purposes rather than for comparing safety records of the airlines of those states.' Changing thinking such that an understanding of the influence of national culture (or even national economic factors) is recognised as being a worthy safety pursuit will not be an easy task for any researcher.

The area of aviation culture that seems to have advanced the most is related to the development of crew resource management (CRM) within the flight deck. A great deal of the early developments in CRM occurred in the US with an apparently positive effect on flying safety. With this initial success as encouragement, certain airlines attempted to sell CRM courses to overseas airlines. One anecdotal source of evidence tells of a group of Korean air crew who were told that CRM required an openness of communication on the flight deck such that even the most junior crew member could speak up if they considered there to be a problem. All of the participants strongly agreed with the sentiments of the course facilitator and then went back to line flying and continued with the belief that the Captain is always right and junior crew members do not question his judgment. In the context of the classroom, the American facilitator held the key position of power and was therefore "always right". Whatever he said about the way the students should behave was agreed with. As soon as the crews were back flying, the Captain held the key power position and went back to being always right and therefore not open to question.

The major players in examining the effects of CRM and in particular how it relates to culture have included the joint NASA / University of Texas (NASA/UT) Aerospace Crew Research Project run under the direction of Bob Helmreich. Their work in the area of aviation human factors has covered nearly 20 years and started with expertise in CRM concepts and training that were specific to the USA (see Merritt, 1993, 1993b, 1997, Helmreich and Wilhelm, 1997). As the profile of the CRM concept has continued to be raised around the world, the NASA/UT team have been able to examine cross-cultural difference in the way people train and react to training. The addition of a cultural dimension to the teams work is a relatively recent phenomena, but one they see as being ever more important. Merritt comments that '...as CRM extends beyond the cockpit and across national boundaries, we are now recognising that the CRM practices which we held to be universally applicable are indeed culturally influenced.'

Johnston (1992) recognises the contribution of culture by stating that 'The social processes which contribute to crew functioning vary within each separate culture and, indeed from organisation to organisation across that culture.' He further suggests that even apparently cultureless phenomena such as the use of standard operational procedures (SOPs) are effected by cultural variations in application by crewmembers. Directing his investigations specifically towards CRM training, Johnston argues that such training '...may be ineffective or rejected in the absence of cultural compatibility'.

Previous to the work of Merritt and Johnston, however, was that of Redding and Ogilvie (1994) at the University of Hong Kong, into the cultural effects on communication in the cockpits of commercial aircraft. The authors suggest that although flying crew carry '...feelings, attitudes, beliefs and values..' which are derived from their individual personalities, it is possible to aggregate these features into larger groups such as national cultures. 'This cultural level of difference, although it is assumed not to operate in affecting professional behaviour of flying crew, may in fact, be operating unconsciously and in ways which are difficult to perceive.'

Much of the focus of Redding and Ogilvie's work is aimed at the impact of hierarchy on the openness of communication, a concept which is now embraced more formally under CRM. Breakdowns in crew performance being distinctly different to individual failings and arguably more consequential. Ashford's study of 219 large aircraft accidents found that *failure to cross-check / co-ordinate* was an accident causal factor in 118 accidents or 54% of the total. (Ashford, 1994)

It was the expert opinion of several aviation safety professionals around the world (Lewis, 1993; Learmount, 1993; Green, 1993) that one of the fundamental strengths of the Australian aviation system is an openness in communication, particularly on the flight deck. Lewis, Head of Safety and Environment for Qantas Airways suggests that, 'Perhaps the element that exists in Australia that is more prevalent than in other cultures, is that Australians respect authority, but do not defer to it. In other words, if the First Officer is not happy with what the Captain is doing, he will tell him so.' The late British aviation psychologist, Roger Green enlarged on this point by stating that '...the nature of crew interactions on the flight deck in Qantas is different from most airlines for two possible reasons.' (The first reason relates to heavy crewing, which is mentioned elsewhere in this text.) The second reason is '...that Aussies are naturally both socially relaxed and disrespectful of authority. The suggested impact of this is that status differences do not count for a lot on the flight deck and captains thus find it easy to make the best use of the other crew members' ideas and F/Os have no inhibitions about letting the captain have the benefit of their own thoughts'. Nevertheless, a note of caution is warranted at this point. The substitution of organisational culture for national culture is a great temptation, especially when considering flag-carriers. In other words, some cultural traits observed at national level in aviation may in fact be organisational cultural traits and therefore deeper investigation by the researcher is required.

The Story so Far

Redding and Ogilvie's (1984) early study of culture and communication examined 151 flightcrew from eight different carriers in an attempt to establish whether

- the attitudes of flight crew reflect their cultures
- the atmosphere on a flight deck reflects a specific culture
- attitudes or flight deck atmosphere effects communications.

Their model of the factors that contribute to the quality and frequency of communications provides an interesting way of illustrating what factors may be relevant in a safety investigation such as this. The authors do not attempt to measure organisational culture, but accept that each airline will inevitably have a different culture.

Figure 6.2   Contributory factors to cockpit communications

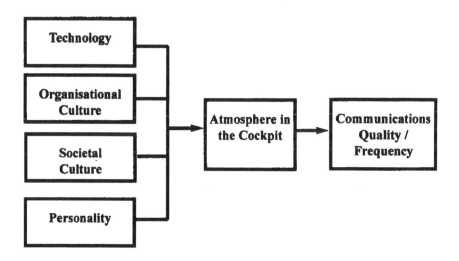

Source: Redrawn from Redding and Ogilvie, 1984.

Redding and Ogilvie convey some frustration in the value of their results, which seems to stem from the small samples available to them. Apart from limitations in the number of flight crew who were prepared to file questionnaire returns, finding enough airlines that fit within certain cultural groupings also seemed to prove difficult.

Nevertheless, the authors conclude that there is some evidence that;

- societal values about the distribution of power are brought into the cockpit
- that they influence the climate in which communications take place
- that the more open flow of communications is conducive to better performance.

In summary, they also warn that these conclusions remain tentative and further research is needed in the area before steadfast conclusions can be drawn.

Johnston (1992) presents the work of Ooi (1991; 1992) in terms of his attempts to simplify some of the classical work done by Hofstede into broad differences between Eastern and Western Cultures. At its simplest level, Ooi summarises the differences as follows;

Table 6.2 Cultural dimensions

Cultural Dimensions	Eastern	Western
Individualism: Collectivism	Collective	Individual
Uncertainty Avoidance	High	Low
Power Distance	High	Low
Masculine: Feminine	Masculine	Feminine

Source: Adapted from Johnston, 1992.

One of the many cultural concerns that these differences present is in what is termed the cockpit power gradient. Similar in nature to Hofstede's concept of power distance, the cockpit power gradient essential concerns itself with the ability of pilots to work as a cohesive crew as oppose to a hierarchy. As flight decks become more multicultural, the power gradients may be expected to become more unpredictable. Already Malaysian Airlines fly with 35% expatriate crew members who represent 16 different countries (Flight International, 1994) and Singapore Airlines fly with 48 different nationalities of crew members. Ooi presents a summary of the differences in the following table.

Table 6.3 Cockpit culture combinations

Possible Combinations	Power Gradient
Western captain / Eastern first officer	flat / steep
Western captain / Western first officer	flat / flat
Eastern captain / Eastern first officer	steep / steep
Eastern captain / Western first officer	steep / flat

Source: Adapted from Flight International, 1994.

At such a high level, significant cultural differences are to be expected. The fact that it is possible to use the terms *Eastern* and *Western* culture without needing to define them too closely is one indicator of how well recognised the differences are. However, the use of such findings at the micro level of training and operations is somewhat limited. There are so many other variables to add to the above equation before it presents any use-able conclusions about the problems of multicultural crewing. For example, in terms of large aircraft manufacturers, there are only a small number of players around the world. Boeing builds with a western (American) cultural backgrounds, Airbus with western (European) philosophies and the likes of Ilyushin, Tupolev and Yakovlev with eastern (Russian) principles. These aircraft types are then flown by a mix of cultures around the world. Therefore although every B747, for example, is essentially built as the same aircraft regardless of whether it is to be flown by a Chinese or American crew or, indeed, a mix of the two, it is difficult to tell how much effect a cockpit gradient is having on different "cultural designs" of aircraft.

Johnston (1992) dares to discuss the effect of regional variations on accident causation, but notes how rarely the subject is broached partly because of '...the absence of research and partly the potential sensitivity of the subject'. He also notes that what little evidence there is in the field is largely unstructured and anecdotal and therefore of little scientific validity. Kenton-Page (1995) of Hughes Flight Training is a Captain with a long his-tory of training UK and foreign nationals. He documents that in training various nationalities, particularly in the area of crew resource management, there are major differences in attitudes. He states that '...some nationalities consider the *Anglo* culture inferior to theirs which makes it very difficult to put across any new concept and especially one that may be in conflict with their culture.' CRM is very much a concept that has developed from Anglo cultures, particularly in the USA, UK and Australia.

The concepts of junior crew members being able to speak up and question the actions of their superiors is difficult enough for some *Anglo* pilots to accept, but in countries with higher power-distance ratios, it is often near to impossible. Kenton-Page describes the 'Father/Son' relation-ship which varies from being a real father and son flying together with the understandable difficulties that such personal relationships can bring, to a cultural, functional 'Father/Son' hierarchy on the flight deck which is evi-dent in, say, the Japanese. He adds that Hughes '...also come across diffi-culties because of a tribal or caste system. This is so ingrained that I feel it may be an intractable problem.' (Kenton-Page, 1995)

Perhaps unwittingly, Kenton-Page seems to hit the nail on the head with his statement regarding ingrained tribal or caste (cultural) traits. To describe them as an intractable problem could well be symptomatic of the 'Anglo' attitude towards the results of decades of anecdotal cultural studies. As mentioned earlier, the differences in culture only become apparent when we look at how different other cultures are in comparison. The temptation is to look at the subject from the point of view that, 'our own culture is the base line, so what is strange about everyone else?' The use of accidents where collectivist Eastern nationals have flown to their deaths as examples of bad CRM for use in American and British training courses has helped propagate the myth that the West is getting things right and the Eastern nationals will have to undergo a massive change to get to *our* level. This assumes one thing - that the Western way of flying aircraft (if it were ever possible to define that) is the only safe way to operate an aircraft.

For example, although the Japanese seem to be recurrently singled out for displaying behavioural traits that seem hazardous to successful CRM, their strict attitude to flight deck operations may actually be some-thing very safe. The sight of Japanese flight crew wearing pure white gloves whilst they fly may give the impression of a highly regimented atmosphere, not conducive to junior crew members questioning their superior's actions. However, the strict adherence to Standard Operating Procedures that such discipline requires is also one of the reasons that Qantas' Head of Safety and Environment (Lewis, 1995) believes has kept Qantas safe.

A quick review of Sears' (1986) analysis of accident causal factors (see Table 2.2) reveals that deviation from basic operational procedures was the number one cause. Provided that the operational procedures are well thought out and sensible, then a culture that engenders strict compliance may be a very good thing. It is when problems start to arise that the deci-sion-making skills and ingenuity of the whole flight crew becomes of prime importance. It is possible that some *poor CRM* accidents in non-Anglo cul-tures are balanced out through a lower *deviations from SOP's* accident rate.

While this example remains largely theoretical, it does highlight the multiple effects of culture upon aircraft operations. The effect of various cultural dimensions should not be considered in a vacuum as the effects are rarely mutually exclusive. This challenges some of the more simplistic approaches to aviation safety and culture, which have risked trivialising what is still a new and potentially very rich vein of knowledge.

In 1994, Phelan (Flight International) presented data from Boeing which illustrated an apparent correlation between accident rate per million departures and Hofstede's Power-distance and Individualism indices. Although no correlation co-efficient or basic statistics were supplied, a trend line indicated a relatively strong correlation. The relationship between power-distance and accident rate suggests that as the relationship between manager and subordinate becomes a more steep gradient, so too the accident rate would increase. This would seem to support the general arguments behind crew resource management training.

The Boeing data also suggested that as individualism increases, so the accident rate diminishes. This may be because of the perceived ability to speak up when things seem wrong, rather than following the collective view. Australia, USA and UK are all countries with a high individualism and low power-distance score.

The risk of this important, yet relatively simplistic article, was of suggesting that the cultural dimension in itself was the reason behind the variety in safety records. Although Phelan, a well respected aviation journalist, did not set out to make such a statement, the limitation of publishing space made it difficult to launch such a fundamental concept to the aviation community at large. A number of letters of complaint ensued including:

Sir - Your article 'Cultivating Safety' (Flight International, 24-30 August, p22) advises caution when reflecting on the relationship between culture and aviation safety. This requires not only an appreciation of national sensitivities, but an awareness of how cultural identity can lead to unwarranted moral evaluation. Unlike the issue of race, differences founded upon culture might appear to be innocuous. To suggest however that Iran is "collectivist" implies that it is too collectivist, otherwise this would hardly be worth mentioning.

This can be used as an implicit criticism, suggesting that other traits which are lacking and could lead to the belief that the remedy lies in promoting specific aspects of Western European behaviour. Becoming captive to a particular theory brings the risk of being blinded to the third element - the causal factor - which is easily overlooked. It could be plausibly argued that it is economic wealth which allows for both individualism and a healthy aviation industry...
(Flight International, 1994a)

The author highlights the sensitivity of culture as a factor in aviation safety discussions and highlights one of the risks of this subject. That is of judging cultural difference in terms of what the host culture does right and what other cultures do wrong. He also touches on another subject that is often related to national differences, but does not strictly come under the classification of culture; that is the effect of economic wealth.

Boeing's attempt to correlate Hofstede's cultural indices accident rate represents a tentative first step. However, the results are rather primitive and seem to lack the necessary academic integrity. Whilst it is difficult to make judgment without sighting the original data, the scattering of datapoints of the two graphs seemed to make the trendlines seem optimistic at best. In another letter to Flight International (1994b),

> ...the graphs of accident rates against "power distance index" and "individualism" do not support the simple proportional relationships indicated by the straight lines drawn through the points. More worrying is the assumption that a correlation implies causality...

This is a fact that cannot help convince either the academic or industrial community and risks doing more harm than good. In the same issue of Flight International, Johnston, one of the most respected experts in this area, warns that

> ...cultural influences and regional safety are sensitive matters. There is virtually no definitive research upon which to base interventions.

Nevertheless, the Boeing study remains ground-breaking and is afforded some integrity by virtue of their international market. Cultural sensitivities are very relevant to a manufacturer of the magnitude of Boeing, which can only lend credibility to their investigations into the influence of culture on safety.

National Cultural Traits of Australia

So far in this book, an overview of national cultural differences that have been observed through empirical research around the world, it is necessary to summarise those cultural traits which may be seen to be *Australian*. This is necessary so that similarities and differences between the national and industry / organisational cultures can be recognised in the next stage of analysis.

The stereotyped Australian is a loud individual who 'is not backward about coming forwards, says what he likes and likes what he says'. The Crocodile Dundee or Sir Les Patterson image is an endearing if not an enduring one.

Trompenaars (1993) asked a series of questions in his research into cultural diversity, which produced information of relevance to the aviation industry. None of the data was collected as particularly aviation-specific, but there is much to be learnt from his observations. The variability of individualism and power-distance are clearly illustrated through the following questions.

For instance, the author examined the effect of national culture on corporate culture in terms of how managers perceived their leaders. This gives an impression of the relative hierarchy and therefore the role that individuals are expected to play. By asking the question; 'What makes a good manager?' Trompenaars plots the percentage of respondents who opted '...to be left alone to get the job done'.

Figure 6.3   What makes a good manager?
Percentage opting to be left alone to get the job done

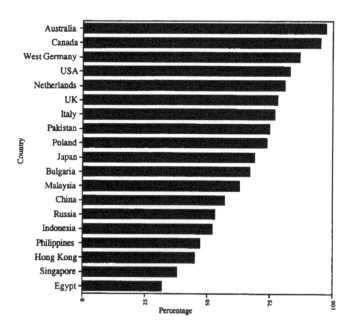

Source: Adapted from Trompenaars, 1993.

Australians firmly believed that the manager was responsible for doing a particular job rather than presiding as a *father-like* figure over his subordinates. In the more familial cultures such as Turkey, Venezuela and China, the role of leaders as '...floating on seas of adoration' (Trompenaars, 1993) and buoyed by the loyalties of their subordinates is clear. It is also reflected in organisational structures, where these countries demonstrate the steepest hierarchies. Australia on the other hand reflects a relatively flat hierarchy.

The importance of the flat hierarchy for aviation safety lies at two levels; namely on the flight deck and within aviation organisations. Both represent similar group dynamics although the former is probably the easiest to recognise as being present. The key issue is one of communications both in terms of integrity and speed. Aviation represents a time critical operation with few opportunities for trial and error. Crew Resource Management highlights three aspects of behaviour that are vital to safe flightdeck operation; namely command, control and communication. The need for effective command dictates that communication gradients should not be perfectly flat (i.e. no leader) and yet the need for open communication requires the command situation not be too much of the other extreme. An excessively steep communication gradient may delay or even prevent the flow of information which may be time or safety critical.

These principles also apply within organisational dynamics. Indeed, just as the original CRM courses derived much of their material from the business world, so the principles that aviation relates to as CRM can be shown to apply in reverse. Of course, that is not to say that organisational processes have been evolved from CRM principles, even though advocates suggest that the next generation of CRM is *corporate* or *company* resource management. Such an effect remains at the planning stage although both Ansett Australia and Qantas intend to introduce company-wide CRM-type training in 2001.

Open communication by itself does not entirely capture the cultural identity that anecdotal evidence attaches to Australia. It is the directness of style that is often focused upon, although it is arguable that open communication within Australia is indeed direct in style. However, as this is not necessarily the case for other cultures where communication channels may be open, albeit with much direct language use, it is worth considering as a separate issue.

Trompenaars (1993) highlights the cultural tolerance to direct communication through a reasonably obscure question. Respondents from 38 countries were asked what they would do when faced with the following problem;

A boss asks a subordinate to help him paint his house. The subordinate, who does not feel like doing it, discusses the situation with a colleague. The colleague argues: 'You don't have to paint it if you don't feel like it. He is your boss at work. Outside he has little authority.' The subordinate argues: 'Despite the fact that I don't feel like it, I will paint it. He is my boss and you can't ignore that outside work either.'

Trompenaars wanted to assess what proportion of the respondents would feel able to refuse to help their boss. The results (shown for a selected 18 countries) were documented as follows;

Figure 6.4   House painting
Percentage who would refuse to help the boss

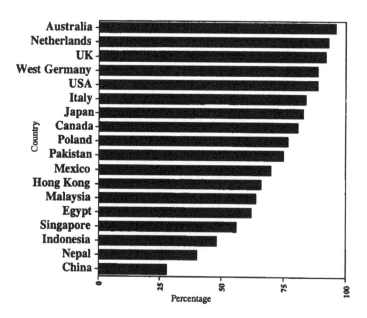

Source: Adapted from Trompenaars, 1993.

Australia ranked the highest (96%) demonstrating not that Australian's were unhelpful, but rather that it was culturally acceptable for them to decline if they really did not want to help. The likelihood that a manager would consider the refusal to be an insult or act of insubordination is low and neither would he suffer from a loss of face. Subsequent discussions with various groups of Australians about this graph reveals that the decision would very much come down to whether the boss was thought to be "a mate" or not. The critical factor not being the ascribed rank or position of the manager, but whether he had earnt the respect and friendship of the subordinate. Knowing that it is acceptable to say no allows the decision to be more directly influenced by "softer" issues such as friendship.

In terms of aviation, the Trompenaars graph may be interpolated to support the premise that 'Australians do not defer to authority' (Green, 1993). That is to say that individuals do not defer to authority for authority's sake, rather they maintain a healthy disrespect for it which allows decisions which are perceived to be fallible to be questioned.

The history of modern Australia spans over 200 years and it is beyond the scope of this text to explore the area in great detail. However, there are a few issues which have stood out as being relevant to flight safety by crafting certain aspects of national culture. This is not to say that the link has been formally established. Indeed as Patience (1991) points out, 'In Australia the recognition of culture as a critical force shaping consciousness and action has not been widespread. It is largely confined to history and remains relatively untheoretical, particularly in sociology which is one discipline one might most expect to have taken up this theme.'

Patience cites Ward's *The Australian Legend* (Ward, 1982) as perhaps the best known historical interpretation of Australian culture. 'This offers a picture of an egalitarian culture shaped by the bush, mateship, self-reliance, toughness practical inventiveness, and coloured by a larrikin irreverence for most authority.' The emphasis appears to be on "mateship" with a commensurate discrimination against "non-mates". The definition of non-mates generally encompasses those who '...do not easily slot into the category of Anglo-Celtic male including Aborigines and women'. Mateship is not just about the "easy-going" and friendly image which many Australians would like to project. Indeed, mateship is very much about groups rather than an "open membership" policy of friendliness. This is possibly a situation compatible with an efficient airline industry and yet one that could also ultimately have a destructive effect.

Historically, as White (1981) writes, 'The emphasis was on masculinity and on masculine friendships and teamwork, on "mateship" in Australia'. The flight decks of multi-crew aircraft also tended to be almost exclusively male domains. A shared love of flying and a cultural context that allowed flight deck crew of different ranks to be mates tended to support the objectives of crew resource management long before anyone coined the phrase. This is a crucial issue that is worth noting; that many aspects of culture exist and have existed well before they have ever been labelled. Indeed some of the resistance to cultural change spans from practitioners who do not recognise the existing culture and attempt to teach "old dogs" new tricks which are in fact old tricks with new names. Australians in particular, renowned for their '...forceful behaviour and drive' (Johnston, 1993), do not generally warm to being taught things they already know. Polite deference to authority in terms of listening to things they already know is certainly not an Australian trait. However, that having been said, it is also an Australian trait to believe in giving people "a fair go". More simply, the attitude seems to be, "listen to what they have to say and if it is rubbish, tell them!"

Exploring the issue of authority and communication

The perceived distance between manager and subordinate in Australia is expected to be low, according to Hofstede's (1980, 1991) and Trompenaars (1993) work and through the anecdotal evidence supplied by expert witnesses.

One way to explore this relationship on the flight deck, particularly in terms of the Captain (manager) and First / Second Office / Flight Engineer (subordinate) relationship was inspired by a study which originated from the RAF Institute of Aviation Medicine. Beaty (1995) and Taylor (1991) both refer to the study of military crews which revealed that the following words were used to describe other crew members. (see table 6.4)

Table 6.4   Descriptions of fellow crew (UK military)

Captains describing co-pilots	Co-pilots describing captains
Competitive	Over-confident
Over-confident	Arrogant
Strong personality	Abrasive
Obstructive	Bad tempered
Obnoxious	Unpleasant
Bolshie	Sarcastic
Difficult	Over-critical
Unco-operative	Not easy to get on with
Bored	Intransigent
Lazy	Pig-headed
Number-chaser	Unpredictable
No sense of humour	Aggressive
Minimiser	Disagreeable
Complainer	Martinet
Lethargic	Tyrannical
Resentful	Autocratic
Bullying	Authoritarian
Talkative	Incompetent
	Overbearing

Source: Adapted from Beaty, 1995 and Taylor, 1991.

As can be seen, the comments represent varying levels of negativity and are used by the authors to highlight the problems of the (generally male) pilot's ego. Whilst only a limited amount of information exists about the original study, it does give a quite negative impression on the relationship between crewmembers. Whilst some would argue that an individual's feeling would not affect their "true professionalism", that assumes that everyone acts professionally all of the time; a fact which is obviously untrue or human factor accidents would never take place. The feelings that an individual has towards superiors, or subordinates, will affect the way they communicate and even perform. Indeed, both factors may also prove to be safety critical.

In the course of this case study, Australian crew members were asked to describe other ranks in a repeat of the RAF study. The objective was to examine the perceived relationship between ranks to explore the power-distance effect. Simple analysis of the words which were used to describe other ranks produced a list similar to the RAF study ranking the following (see table 6.5).

Table 6.5   Descriptions of fellow crew (Australian civil and military)

Top 20 words to describe Captains	Top 20 words to describe other crew members
Professional	Professional
Experienced	Competent
Competent	Keen
Knowledgeable	Friendly
Skilled	Helpful
Capable	Supportive
Helpful	Knowledgeable
Approachable	Enthusiastic
Confident	Capable
Friendly	Eager
Responsible	Conscientious
Dedicated	Willing
Managers	Skilled
Enthusiastic	Dedicated
Leaders	Responsible
Conscientious	Experienced
Fair	Inexperienced
Reliable	Reliable
Diligent	Young
Keen	Motivated

In the case of Captains, all of the top twenty words may be considered to be positive, describing as they do both aptitude and attitude. Certainly most of the terms describe features that are desirable in aircraft Captains such as experience, knowledge, capability and aspects of leadership. Approachability, helpfulness and fairness are also traits which appear to concur with the anecdotal evidence regarding Australian aviation and national cultures.

The words used to describe junior crewmembers are also ostensibly positive. Competency, enthusiasm and the ability to support are all traits that would be encouraged by a good commander. The inclusion of *inexperienced* and *young* as the closest the top twenty came to negativity may simply reflect an objective description rather than implying a deficiency.

Words used to describe management may be expected to be rather more negative than those towards co-workers, as is the nature of the job. However 12 of the top 20 words may be seen to be positive and only four were truly negative. The term *professional* ranked as the top description for every position, which is a reflection of the perceived function of all involved in flight operations. (In describing themselves, pilots were not expected to be unduly modest which would explain why all of the top 20 descriptions were positive. Words focus on ability (professional, competent, capable, diligent, safe *etc.) or on attitude (dedicated, conscientious, approachable and enthusiastic).)

Myths of National Culture

Observing differences in national culture is an activity which is prone to significant unscientific biases. These may range from lighthearted stereotyping to rather more serious racism. While there may be some truth behind these anecdotal perceptions, their value in scientific study should be observed with some caution. Stereotyping may be based on personal experience (which in turn may be somewhat limited) or even a projected image from the media and can be a source of both entertainment and misinterpretation.

Australian icons tend to be based on a romantic bushman image propagated through a number of movies. The most successful must be Paul Hogan's *Crocodile Dundee* who delighted audiences with his shrewd, yet seemingly innocent view of the world. On arrival in New York, Mick Dundee is observed riding in the back of a taxi which stops at a set of traffic lights. At this point the hero leans out of the car and introduces himself to two men who are stood chatting at the side of the road; 'G'Day; Mick Dundee from Australia - I am in town for a few days; I'll probably see you around?' The two men look at each other suitably bemused, enhancing the image of Australia as a small country where everyone knows everyone.

However, even this incident has some aspects pertinent to aviation safety. Australia is a small community especially when compared to the USA or Europe. There are a relatively small number of key players in the aviation industry, which can either be a good thing (in terms of close communication) or a bad one in terms of the strength of adverse individual influence or lack of experience.

Numerous Australian films have painted a picture of the strength of the "genuine Aussie battler" and their fight against the system. Ward's (1982) '...larrikan irreverence to authority' is clearly demonstrated in films such as *Strictly Ballroom, Muriel's Wedding* and *The Castle*. It is an unsurprisingly romantic image and has as much to do with the way Australia would like others to perceive it to be, as with the truth. This has helped create an image of a small-town mentality which tends to ignore some of the great innovative achievement of the Australian nation.

Qantas - the American Airline?

In the recession of the early 1990's when Domestic regulation was being reformed in Australia and Qantas was preparing to become a public company, a movement began to review whether the airline should undergo a name change. Although the word Qantas (Queensland and Northern Territory Aerial Services) was, for many, synonymous with Australia, there were also some who believed that the name confused people.

A survey of customer attitudes was conducted both within Australia and overseas which uncovered a significant yet surprising fact from the USA. That was that not only were the American public fiercely patriotic towards their own airlines, but that a significant proportion believed Qantas to be an American airline. The conclusion was that a name change would risk a substantial loss in American traffic rather than any increase.

Australia's view of the world

The other aspect of cultural stereotyping is that it works both ways. Whatever the image different countries have of Australia, there are similarly interesting perceptions held by the Australians. The impression of straight-laced Brits which seems to be summed up by actors such as Anthony Hopkins, Laurence Olivier, or Hugh Grant sits alongside images of Americans as hard-talking individuals like John Wayne, Tom Cruise or

Harrison Ford. That having been said, the multicultural nature of Australia, and particularly its aviation industry reduces the effect of a lack of experience. Expertise within the industry is drawn not only from Australians, but a large number of expatriates. Although Qantas and Ansett will only recruit staff with Australian citizenship, they have encouraged and assisted international staff, especially flight crew, to relocate and naturalise.

Whilst it has been suggested that culture can remain invisible until it is compared to another, it must also be noted that perceived differences can be based on stereotypes, or at least limited exposure. As a consequence it is easy to perceive cultural superiority, or indeed be led to believe a cultural inferiority. In the former case, it is tempting to view developing nation's safety records with some contempt and risk ignoring the fact that similar basic issues may exist as latent conditions in both aviation systems. For example, the VH-INH B747 accident at Sydney Airport (BASI, 1996) was partly the result of very basic issues, namely training and communication. In the latter case, of cultural inferiority, there seems to be a mood within Australia to look towards the USA and Europe for expertise, a movement which is not always justified. Whilst CASA has made a commitment to harmonising its regulations with the American FARs and European JARs, some caution is warranted for fear of the "grass is always greener" factor. Former CASA Director, Keith (1997) warns that '...in my opinion, the FAA today is in a distressed state, not dissimilar to the Australian regulator in the 1990s.'

Organisational Culture

Reason (1993) defines corporate culture as '...the set of unwritten rules that govern acceptable behaviour within and outside the organisation. It emanates from the strategic apex of the company and colours all of its activities'. Mitroff and Kilman (1984) define organisational culture as '...a set of shared philosophies, ideologies, values, beliefs, expectations, attitudes, assumptions, and norms', whereas Jackson (1960) more simply states that 'cultural norms refer to the set of unwritten rules that guide behaviour' which is closer to Reason's definition. Every organisation has its own unique culture and even sections of the same company may have specific traits or ways of operating. At its simplest, organisational culture is concerned with "the way we do things around here". Kroeber (1952) makes the point that'...culture is learned and transmitted through groups and individuals in society'. In other words, organisational culture is self perpetuating and not solely dependent on the individuals who form that organisation.

According to Trompenaars (1993), 'the organisation... is a subjective construct and its employees will give meaning to their environment based on their own particular cultural programming.' The culture of that organisation is '...shaped not only by technologies and markets, but by the cultural preferences of leaders and employees.' Certainly in a new company or one undergoing significant high level restructuring, the key employees will adopt traditions or ways of behaving that are already familiar to them. Mental images of 'this is the way we did things at company x' or 'this is the way we should have done things at company x' will have a strong influence.

Organisational Culture and Aviation

Lauber (1993) examined a series of high technology accidents for evidence of an influence from organisational culture. He cites the examples of the Ro-Ro ferry, *Herald of Free Enterprise*, the Clapham Junction railway disaster and the loss of an Embraer 120 near Eagle Lake, Texas. The legal action against P & O management as a result of the 1987 *Herald of Free Enterprise* disaster broke new ground when it was ruled that a body corporate could be responsible for manslaughter. This came when perhaps there were employees who felt that by placing the blame upon the assistant bosun, they were safe from accountability. Fortunately, the Inquiry looked beyond the active failure of the mariner, who was asleep in his cabin when he should have closed the bow doors, and examined the latent defects that allowed such an error to lead to a disaster. The Public Inquiry had been most explicit in detailing where the fault really lay; 'All concerned in management, from the members of the Board of Directors down to the junior superintendents, were guilty of fault in that all must be regarded as sharing responsibility for the failure of management. From top to bottom, the body corporate was infected with the disease of sloppiness.' (DOT, 1987) It is most unlikely that a collection of faulted individuals came together by chance. Similar attitudes to safety within an organisation reflect the organisational culture and is perpetuated through the subtle influence of "how we do things around here" or in the recruitment of subordinates in the image of their superiors.

Safety Culture

A review of safety and organisational culture literature, will find this term used quite commonly. Meshkati (1997) contends that it was first introduced following the Chernobyl nuclear power plant accident and is used extensively by a number of industries including chemical processing. Reason (1997) points out that 'Few phrases are so widely used yet so hard to define as "safety culture".' He further explains that '...there is nothing mystical about it. Nor is it a single entity. It is made up of a number of interacting elements that have enhanced "safety health" as their natural by-product.'

It is important that there is some sort of definition to establish whether there is a real difference between a "safety culture" and a "safe culture". As Wood (1993) puts it, 'Is it our corporate attitude toward safety or is it the safety program itself?'. Should the concept of safety culture be kept separate from the overall culture of an organisation and if so, is this not a self-defeating definition? That is to say, is it possible for an organisation's overall culture to be considered healthy even its safety culture is not and vice versa? Indeed, closer investigation of work on safety culture suggest that the authors are in fact examining the safety focussed aspects of an organisation. The danger here being that aspects of corporate operations that may in fact have safety consequences may not be perceived to do so unless they fall under the more obvious definitions of "safety programs".

Westrum (1997) writes that 'Around every technological system there is a human envelope of protection. This envelope is made up of those who originate, operate, and maintain the system, and together they form the protective elements that keep the system intact and safe from harm. This envelope may be thick or thin, seamless or faulted, but this envelope protects the system from harm.' While he avoids referring to this "human envelope" as the safety culture of an organisation, it represents the meeting of organisational and safety cultures. Maurino (1992) suggests that '...the design and corporate culture of an organisation exert powerful influences on how safely it functions. Pilots, controllers and other operational personnel do not act in a vacuum - instead they mirror the policies and practices of the organisations to which they belong.'

For example, the subject of commercial pressure in placing unreasonable demands on operating staff has started to have real safety consequences in air transport operations. In the context of the Australian domestic airline market this has been demonstrated where the commercial sides of the airlines battle for customers they have started to offer more and more in the way of service add-ons. Even on short domestic sectors, business and first class passengers have been led to expect their coats to be stowed by the flight attendant, pre-take off drinks, a choice of newspapers, and an on-time departure. On busy flights or those attempting to catch up time from an inbound delay, flight attendants have been known to be still carrying on their duties in the cabin during the take off run. This has now become a safety issue, but the systemic problem began as a corporate attitude (priorities of customer service) and became a "safety culture" issue at a later date.

Wood (1993) describes organisations with strong safety programs as displaying seven basic cultural traits. The most important of these being the involvement of management from the top. This is, of course, a basic premise to the success of most organisational operations and not exclusively the domain of safety programs. The concept of Total Quality Management (TQM), which has been adopted with great success, particularly in the freight logistics industry, relies very heavily on this sort of management, yet can hardly be described as a safety program: Or could it? If the efficiencies which are strived for in TQM are really met then they will have a commensurate safety benefit - good safety is essentially the same as operational efficiency.

Doak (1993) describes American Express' corporate aviation safety culture as being broken down into five areas;

- Beliefs
- Corporate attitude
- What was in place
- Plan
- Action.

Doak also describes how the '...safety consciousness existed at the highest level in the corporation, including the chairman.' But is this really an element of a safety culture? Surely this is a confusing expedition into semantics? A safety culture is not a mutually exclusive component of an organisations culture; it is an important descriptor of one of its attitudes.

O'Leary and Pidgeon (1995) suggest that a '...good safety culture is characterised in four ways;

1. It originates at the level of strategic management.
2. Concern for safety is distributed and endorsed throughout the organisation.
3. There is a clear set of flexible and effective norms and rules which govern safety behaviour.
4. There is ongoing pro-active reflection on unsafe events and incidents and about safety in general.'

O'Leary and Pidgeon also raise an interesting idea for assessing the safety culture of an airline through the company's attitude towards incident reporting. In addressing a comment from ICAO's Stephan Corrie that 'It's too bad that we need to have confidential reporting programmes', the authors have interpreted the statement to mean; '...in an ideal world, reporting systems would exist within a culture in which line pilots would be able to discuss their technical, operational, crew and personal problems directly with their managers.' In other words, the fact that airlines have to set up confidential systems is symptomatic of a level of distrust between employees and management, which may restrict the flow of safety information. It also causes problems in that management may be reluctant to accept and act on information from anonymous sources, which cannot be verified. (In some cases this can also be abused as an excuse for not acting on a report).

If communication lines were open enough for problems to be reported without the need for confidentiality, then the time needed for safety problems to be resolved can be reduced and the number of reports may increase. Indeed, both British Airways (Wright, 1994) and Airservices Australia (Guselli, 1995) highlighted with pride the increased number of incident reports within their respective organisations. They are not the result of an absolute increase in reportable incidents. Instead they reflect an increasing level of trust in the system as a method of communication and the ease of reporting.

The argument against the existence of a safety culture separate to that of corporate or organisational culture is based upon the difficulty of delineating between safety and non-safety actions. Indeed it is this difficulty that can be found to be the critical error in decision making in many accidents. When decisions are known to have a direct impact on safety, the choice process works quite differently to when they are not known to have a direct effect. It is often only in retrospect or during formal accident investigation that safety critical actions or decisions are highlighted as such. Attempts to separate safety culture from organisational culture are difficult enough following an incident and therefore are all but doomed to failure in the predictive sense.

Using the four aspect of safety culture as highlighted by O'Leary and Pidgeon (1995), it is possible to demonstrate just how similar it is to organisational culture;

1.   It originates at the level of strategic management:- Organisational culture is also highly dependent on leadership. Cultural change originates from strategic management through policy decisions and support (or otherwise) for particular projects. Charismatic leadership from individuals such as Steve Jobs (Apple), Richard Branson (Virgin) and Bill Gates (Microsoft) can define the culture of organisations in both a positive and negative sense. An inefficient and unhappy organisation is one which will contain latent defects; the precursors for all accidents.

2.   Concern for safety is distributed and endorsed throughout the organisation:- Although this statement appears to be specific to safety culture, a second glance will pose the question as to what constitutes "concern for safety". Whilst explicit safety concerns such as Occupational Health and Safety posters, training programs and audits may be the most obvious concerns, there are countless areas which impact upon safety which are not directly obvious. A good example is in the area of excess baggage, where check-in agents are often encouraged to turn the other cheek to weights above the limit (20kg for economy and 30kg for business class), especially for Frequent Flyer or business class passengers. This has become institutionalised and now passengers expect the extra, which has a potential follow on affect on ground handling staff (Occupational Health and Safety) and aircraft handling (Flight Safety).

3.   There is a clear set of flexible and effective norms and rules which govern safety behaviour:- Once again, although this refers specifically to safety behaviour and therefore apparently suggests the difference between safety culture and corporate culture to be their remit, the problem is one of defining boundaries. Attempting to separate norms and rules which govern safety is a near impossible task not least because apparently minor or insignificant actions can interact with a complex series of factors to cause accidents. Remembering that accidents are, by definition, unexpected events, part of the danger of trying to highlight safety behaviour rules is not anticipating the compound effect. Actions which are safety critical to one employee, may not appear to be so for others. For example, a flight attendant may not realise that the flight deck crew should not be disturbed part way through a checklist, even on the ground after a flight, until they are made aware.

4.   There is ongoing pro-active reflection on unsafe events and incidents and about safety in general:- The follow up to incidents is another apparently safety specific action, but as such, it is not so different from standard commercial practice of review. A proactive response to all incidents, whether they be operational, safety, financial or even disciplinary is as much a function of good organisational procedure as it is a specific safety concern.

Although the O'Leary and Pidgeon definition is safety specific, all of the four elements also relate to organisational culture which is ultimately responsible for the portion that is referred to as safety culture. However, to separate the two is to make definitive judgments about what is and what is not a safety issue and this is the source of many safety problems in the past. 'Safety is influenced by culture, which dictates priorities, and by pride. The culture can partly be national - affected by the respect in which aviation is held in the culture.' (Flight International, 1994d) In other words, safety isn't a culture in itself, rather it is a condition defined by multiple levels of culture. If the concept of a safety culture is something which focuses the corporate mind towards the importance of safety, then it is a good thing. However, if there is a risk that it separates the issue of safety from general operation efficiency, then it should be discouraged as they are not mutually exclusive from each other.

Industry Culture

The existence of national and organisational cultures seems to make the existence of an industry culture a likelihood, although there are few studies that have ever tried to examine this mid-range. Traditional arguments against such a phenomenon include the apparently transferable skills of managers who seem equally able to work in say manufacturing, retail and transportation. Whether this is true or not is the source of some debate and may actually be the first pointer to the existence of *professional* cultures which hold together industries to create a *virtual* industry culture.

For instance, there are professional qualifications for pilots, maintenance engineers, and accountants that are identical whether the employee belongs to Qantas or Ansett, but then in the case of the latter category, the skills would be similar whether they worked for an airline, a manufacturer or a retailer. Nevertheless, there are certain traits even within that job that would make an airline accountant different to a non-airline accountant.

Defining aviation industry culture is a difficult job and one that has not been achieved in any form other than anecdote. Yet it remains a worthwhile exercise to attempt. Differences in organisational, professional, industry and national cultures seem to have a significant effect on safety; a fact reflected in the more recent focus on safety culture as a safety management strategy. It is a task that requires significant experience of more than one industry and more than one organisation within each industry. It is also a task that would require multinational collaboration and herein lies the first challenge. A research group would need to conquer its own cultural differences before being able to objectively examine other aspects of culture.

Perhaps the first step is to explore the differences between aviation and other transport modes to define what makes it different. Culture is best seen in the context of other cultures and this is the challenge for chapter 7.

7 Comparative Safety

The Relative Safety of Aviation

In exploring the relationship between culture and safety within the aviation industry, is important to try and establish whether any phenomenon is the result of a national, industry and organisational culture, or indeed a mix of all three. One way of doing this is to ask the question, *Is Australian airline safety above average because of a safe national culture?* In other words, do Australians have an inherent aversion to risk and therefore a low level of risk acceptability?

Ideally, the researcher requires a set of activities which are similar in nature and measurement between countries for which there is readily available data for a stable period of time. Although various measures of risk exist, the use of fatalities as a measure is the most sensible for analysis. A fatality is a strict definition which is easily recorded - levels of injury are difficult to define and open to subjective interpretation.

The transport industry provides a number of readily published statistics of accident rates, especially for car travel. Solomon (1993) published fatal accident rates for 26 countries which demonstrate considerable variation between countries, especially when normalised against the number of registered cars. His data is reproduced in Figure 7.1 overleaf. Unfortunately, a complete set of data for more countries is not available and this means that it is difficult to try and attach any cultural significance to the results. Whilst in terms of road safety a country like Egypt looks to be most unsafe, if statistics for general accident rates are shown, it appears relatively safe (see figure 7.2).

Australia does not stand out as being especially safe and is situated amongst relatively *similar* nations, both in terms of economic status and culture. It is certainly not a particularly unsafe nation in terms of accidental death rate, especially compared to some of the more developing industrial nations such as Hungary and Cuba. However, whilst the Figures 7.1 and 7.2 may be used as a guide, they are not necessarily 100% accurate as the *process* for recording an accidental death may vary between nations.

Figure 7.1   Automotive deaths for 26 countries

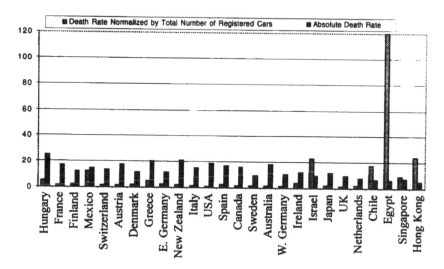

Source: Redrawn from Solomon, 1993.

Figure 7.2   Accidental deaths per 100,000 people per year

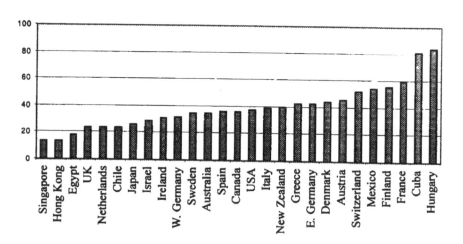

Source: Redrawn from Solomon, 1993.

A study of road accident fatality rates conducted by the Transport Accident Commission (TAC, 1995) of Victoria, Australia looked at accident rates for eight countries. The results placed Australia as shown in Table 7.1 below.

Table 7.1   Road fatality rates for 8 countries (1993 data)

Country	Fatalities per 10,000 vehicles	Fatalities per 10,000 population
Australia	1.8	10.9
Canada	2.0	12.3
Germany	2.1	12.3
Japan	1.6	8.8
New Zealand	2.7	17.1
Sweden	1.6	7.2
United Kingdom	1.6	6.8
USA	2.0	15.6

Source: Redrawn from TAC, 1995.

The Australian rates are relatively unremarkable compared to the other seven countries. Fatalities per 10,000 vehicles (1.8) are marginally worse than the United Kingdom (1993) and marginally better than the USA (2.0). In terms of fatalities per 10,000 population, Australia (10.9) fares somewhat worse than the UK (6.8), but significantly better than the USA (15.6) and New Zealand (17.1). The reasons behind this are complex and involved (including quality of infrastructure, terrain, speed limits etc.) and beyond the remit of this book. However, road safety statistics rate amongst some of the most easily comparable statistics and therefore help to illustrate a simple point. That is, that Australia in general does not stand out as a particularly *safe* nation, either in terms of road safety or in accidental deaths cumulatively. Neither is it a particularly dangerous nation, especially when compared on a World scale.

Concentrating on accidental deaths within Australia; a difference in attitude to death by various causes may give some indication as to whether safety-consciousness is a cultural trait. Lane (1964) examines the value of life as determined by society through a number of hazards and the resources allocated to counteract them.

Table 7.2 Hazards and countermeasure costs

Hazard	Countermeasure	Use	Cost (£)
Poliomyelitis	free immunisation	in force since 1956	8750
death by accident to motorcyclists	compulsory wearing of crash helmets	in force in Victoria since 1961	3200
death by accident in particular type of training aircraft	compulsory modification to aircraft of this type	rejected proposal	4300
death by accident to car occupants	compulsory fitting of seat belts	legislation rejected in one state	8000
death by accident to road users	improved street lighting	piecemeal application	8000
pulmonary tuberculosis	case finding, free treatment, financial support for sufferer and family	in force	13,800
death by shark attack	reduction in shark population by regular meshing	in force at Sydney beaches since 1937, elsewhere recently	14,000
passenger death in airliner accidents	aft-facing passenger seats	rejected by industry	2.8×10^5
death due to burning in airliner accidents	automatic fire-inerting of engine nacelles	rejected by industry	7.6×10^5
death by drowning of passengers of airline coming down in the sea	carrying of dinghies etc. on overwater flights	standard practice	1×10^6
death by burning in airliner accidents close to airport	full-time firecrew at airports	current Australian practice	3×10^6

Source: Adapted from Lane, 1964.

Eleven events are compared on the grounds of the hazard, counter-measure (and its use) and cost per life saved. Two of the hazards are diseases (polio and tuberculosis) and five are related to aviation accidents. The rest are other accident types. Although the figures are now over thirty years old, they provide a useful historical indicator of the relative costs and allocation of resources in Australia at that time.

Lane's figures (see Figure 7.2) demonstrate the diversity of costs of life that exist even within a single nation. The spectrum of values is from £3,200 to £3,000,000 (1964 costs); a range of one thousand fold. This provides a graphic demonstration of the fact that society attaches different values to death depending on the method rather than the outcome. The author comments that; 'It is apparently acceptable to be drowned but bad to be eaten by a shark; its apparently acceptable to be killed in a bus accident but bad to be killed in an plane accident; it is bad to be crippled by poliomyelitis, but in order if the crippling is caused by an automobile.' Further observation by Lane (1973) notes that, 'tractors in this country kill about three times the number of people as are lost in civil aviation accidents, by a mechanism which has been shown to be almost totally preventable, but no-one takes much notice.'

Aviation, it seems, attracts a disproportionately large amount of expenditure on safety measures which may only save a comparatively small number of people. Not only that, but there is also a clear difference in the acceptability of various causes of death. Whilst no accidental death may be considered acceptable in moral terms, there is a point where the balance between risk exposure and countermeasures are balanced by socio-economic forces. This balance is, in reality, rarely zero, but neither is it the same for different causes. The Air Safety Regulation Review Task Force (1990) also comments on the considerable disparity in risk acceptability between road and air transport within Australia.

The fact that the Australian community continues to accept a very high road toll without undue protest is implicit acceptance of the number of fatalities and the associated cost. Statistics published by the Bureau of Transport and Communications Economics for 1988 show fatalities at 2,886 at a cost of $1,382 million (AU) or $479,000 per fatality. Total costs including injured medical services, legal, insurance, police and vehicle damage totalled $6,180 million. Perhaps public acceptance is related to the fact that total Australian fatalities of 2,886 in 1988 is only marginally higher than the 1960 figure of 2,605?

Lane (1994) notes the significant drop in accident rates for various transportation methods between the 1960's and 1988 in Australia. In doing so, he also presents the significant difference in fatal accident rates between each mode;

Table 7.3 Fatality rate per 100m occupant-kms

Mode	1960's Rate	1988 Rate
Airline	0.06	0.00
Bus	0.33	0.04
Passenger Car	1.81	0.86
General Aviation	10	3.57
Motorcycle	22	13.75

Source: Adapted from Lane 1988.

All of the accident rates have improved over the period (a function of improved technology), but the level of improvement is disparate, as is the *acceptable* accident rates which appear to have been set by society. The *Airline* accident rate was already low (0.06) and has been reduced to zero (albeit based on a small statistical sample). The 1988 car fatality rate (0.86) is only 47.5% of the 1960's figure (1.81) and the 1988 General Aviation figure (3.57) is only 35.7% of the 1960's level (10). Motorcycle fatalities remained very high in 1988 (13.75), only a 37.5% drop from the 1960's figure (22). It is interesting to note that not only have all the rates reduced, but that the ranking of fatality rates has remained the same. Motorcycle riding remains the most dangerous transportation method, followed by General Aviation, then passenger car travel.

Notwithstanding this, caution is required when comparing safety statistics between modes of transport. Lane's chosen measure of 100 million passenger kilometres is valid and allows easy longitudinal analysis of changes over time within each mode. However, it is easy to get a false perception of relative risk as each mode has a different profile; in terms of journey length, frequency, distance. Accident risk for aircraft, for example focuses primarily around the take off and landing phase so long cruise phases, especially by modern aircraft such as the B747-400 and A340 tends to smooth aircraft accident rates when scored against time or kilometres. At the other end of the spectrum, transport modes such as cars and motorcycles may travel comparatively short distances (in kms or time) yet are exposed to an equally high risk throughout the journey. This can make some accident rates seem extraordinarily high when measured with large units such as 100 million kilometres or hours.

There are a number of theories as to why commercial aviation may be seen to have a *disproportionately* low accident rate, or indeed other transport modes a comparatively high accident rate. Smith (1992) examines the way that the media treat aviation safety suggesting it to be '...subjective and emotive... probably because their prime objective... is to make money for shareholders.' Noting that 20 deaths on the roads over a weekend would hardly rate a mention in the media, the author highlights the fact that in the last 20 years, 60,000 Australians have died on the roads and yet no-one has ever died in a commercial jet airliner. A significant point may be that if accident rates were measured in journeys rather than absolute numbers of accidents, the perception may be rather different. Indeed in other arenas that is exactly what happens. The accident rate for the commercial jet aircraft fleet appears to have been stable since the early 1970's which, whilst not ideal, appears to demonstrate stability. However, if this rate were expressed solely in terms of the absolute number of accidents or the absolute number of fatalities, the picture may seem much more gloomy.

The media's focus seems based on the impact and visibility of large scale accidents and the apparently widespread morbid fascination which sell newspapers. However, Smith (1992) also suggests that one of the reasons for *hype* surrounding air safety is because of the staff working in aviation themselves. Another explanation for the hype is that some of those who earn a living from aviation, including pilots, air traffic staff and maintenance personnel, have long realised that an emotive statement to the media in relation to air safety is often the best way of ensuring high salaries and the continuation of inefficient practices.

Smith's rather inflammatory remark places the blame for '...emotive reaction to aviation safety' on the world aviation industry for failing to adequately communicate the risk effectively. The argument is simplistic and does not adequately address what is a complex issue. In an address to the 1997 Risk Engineering Society Conference, Smith blamed 'excessive expenditure on safety at the Lucas Heights nuclear research facility' on the inability of their risk engineers to adequately communicate the true risk of operation. This, according Smith, had led to the deaths of numerous Australians as nuclear power had taken more than its fair share of the *safety dollar*.

In doing so, Smith seems to underestimate the forces of the public and government. Whilst public perception of safety is influenced heavily by the media, it is also influenced by a series of other factors which are discussed below.

Risk Perception and Acceptability

The risk-taking behaviour of an individual is highly dependent upon a unique and, arguably, dynamic set of factors. Put rather more simply, 'People respond to the hazards they perceive.' (Slovic, Fischhoff and Lichtenstein, 1980.) While the outcome of any risk taking decision is not directly related to an individual's perceived risk, it is this perception on which decision making is based. Indeed, even at policy making level, the meeting of perceptions have an extremely powerful impact on the result. For example, objections to the building of nuclear facilities tend to be disproportionate if solely compared to the hazards calculated through quantitative risk assessment (QRA). It is the fear and dread of catastrophic events that has a significant effect on the perception of risk. 'If ...perceptions are faulty, efforts at public and environmental protection are likely to be misdirected' (Slovic et al. 1980).

Although Slovic *et al.* tend to highlight the negative aspect of the above statement i.e. where faulty perceptions lead to an economic inefficiency, it may be that an overestimated perception of risk has in fact afforded a greater margin of safety and therefore contributed to a good safety record. In other words, has a pessimistic perception of risk led to overcompensation in safety measures with a commensurate higher level of assured safety?

Slovic, Fischhoff and Lichtenstein's deep analysis of risk perception lays a useful foundation for examining the subject. The authors suggest a number of judgmental (heuristic) biases which can affect the individual's risk evaluation process. The include the following;

Availability As frequently occurring events are easier to imagine, less frequent events tend to suffer from inaccurate recollection. This is a process which can be further effected by publicity, such as the fear of shark attacks following the release of the *Jaws* movies or a heightened awareness of aviation safety following a high profile air disaster.

Slovic *et al.* concluded that in general '...rare causes of death were overestimated and common causes of death were underestimated. A study of 41 causes of death revealed that accidental death causes were also overestimated, as shown in the following table;

Table 7.4 Estimates of death causes

Most Overestimated	Most Underestimated
All accidents	Smallpox vaccination
Motor vehicle accidents	Diabetes
Tornadoes	Stomach cancer
Flood	Lightning
Botulism	Stroke
All cancer	Tuberculosis
Fire and flames	Asthma
Venomous bite or sting	Emphysema
Homicide	

Source: Adapted from Slovic *et al.* 1980.

Overestimated events tended to be '...dramatic and spectacular' whereas '...unspectacular events which claim one victim at a time and are common in non-fatal form' tended to be underestimated. Aviation accidents involving large commercial aircraft tend to be both dramatic and spectacular and result in a large loss of life in a single event. Accidents which fall into this category are rarely non-fatal events.

Overconfidence A function of the process of heuristics is a generally misplaced confidence in judgments based upon them. Slovic et al. (1980) assign this to '...people's insensitivity to the tenuousness of the assumptions on which judgments are based.' There is a tendency by convinced more by the medium of communication of risk (e.g. through media impact) than the information which is contained. Personal experience can heighten awareness of specific risks at the expense of skewing judgment on other risks.

Overconfidence can also arise where experts underestimate risks, especially in complex or high technology systems or where effects are chronic or cumulative. For example, the designers of the DC10 underestimated the effect of the rear cargo-door blowing off on the flight controls of the aircraft. The trust in the aircraft on account of the Douglas reputation and regulatory certification was sufficiently high that the aircraft was bought in large numbers. Confidence in the design was knocked severely following several accidents.

Desire for Certainty All technologies involve a certain gamble and their relative attractiveness is a function of the possible gains and losses associated with them. Ways of confronting this uncertainty include either denial (a source of overconfidence) or to '...outlaw the risk'. An example of the former may be car travel where in spite of accidents statistics, which confirm its relative unsafety as a transportation mode, perception of the risk is low. An example of the former is nuclear power where perceived risk is so high that denial is not possible and risk-taking attitude tends to be more directed towards its removal.

A desire for certainty may also be affected by cultural factors, as represented by Hofstede's (1980, 1991) *Uncertainty Avoidance Index*. The two areas of research have not been linked to date and would make an interesting future research project. A desire for certainty in risk taking seems likely to be based on a more general attitude towards uncertainty avoidance.

It Won't Happen to Me A function of overconfidence, which is demonstrated in attitudes to driving (Svenson, 1979) and other risks. Direct personal experience which has been mishap free, supplemented by media exposure to others who have suffered accidents has the effect reassuring individuals that they are safer because of their own exceptional skill. As this particularly applies to driving, where the individual is under control, it also applies in reverse to flying as a passenger in a commercial airlines where as a passive participant, *it might just happen to me*. Arguably the level of skill is reversed; car driving requires a single simple test whereas aviation requires recurrent training and checking.

Boeing 737 G-OBME - A Case Study in Risk Taking

To suggest that the majority of air transport accidents have been the result of *faulty* risk perceptions or poor risk-taking decisions may seem to be a somewhat inflammatory remark at first glance. Surely no-one ever *takes risks* in safety critical situations? But of course, individuals do take such risk-taking decisions as a matter of routine.

For example, in January 1990, a British Midland B737-400 crashed at Kegworth, UK. The aircraft had developed an in-flight defect in one of its two engines and the crew shut down what they thought was the damaged engine. In fact they had selected the wrong engine. A diversion was made to East Midlands Airport (the airline's maintenance base) and on final approach, the single remaining running engine broke up. The aircraft lost altitude and collided with terrain short of the runway threshold, with the loss of 57 souls. The risk-taking decisions which may now be seen to have been faulty in that incident included

Risk taking decision	Made by?
• Are two-engined aircraft safe for passenger transport?	regulator
• Is the 737-400 a reliable and safe aircraft?	regulator / airline
• Were the crew adequate trained to be released to the line?	airline
• Were the crew confident enough in their ability to detect faults?	crew
• Was the aircraft warning system adequate for fault detection?	manufacturer
• Was the engine of adequate design?	manufacturer
• Did the errant engine need to be shut down immediately?	crew
• Should the crew have made a visual inspection in flight?	crew
• Was a diversion to the nearest airfield (Birmingham) necessary?	crew
• Should cabin crew have questioned the PA announcement?	crew
• Should passengers have questioned the PA announcement?	passengers

Some of these risk taking decisions represent a simplification of complex procedures, such as aircraft certification. For example, the Boeing 737 had been certified by a number of airworthiness authorities such as the American FAA and UK CAA. These processes involve a complex series of physical tests including destructive and non-destructive testing of materials and flight tests where aircraft performance is measured against a series of predetermined criteria.

However, even this process can make an ill-judged final decision in deciding *is the risk of allowing this model of aircraft to fly acceptable?* In the case of the original Comet series aircraft, this decision was found to be fallible when two airframes were lost in mysterious circumstances. When a cause could not be found, the UK CAA took the step of revoking the aircraft's certificate of airworthiness until such time as a cause became apparent. The eventual discovery of metal fatigue, which had been effected by the use of square windows in a high speed pressurised fuselage, represented an area of aeronautical engineering without expertise. The deficiencies in the certification process were on account of a lack of knowledge rather than an error or mistake. This illustrates an important facet of risk taking - that is the concept of the unknown, which is particularly important in fast developing technologies such as aviation.

Other decisions in the above scenario represent active risk-taking decisions such as *Were the crew confident enough in their ability to detect faults?* and *Was a diversion to the nearest airfield (Birmingham) necessary?*. The flight crew had recently completed a conversion course from the Boeing 737-300 series (electromechanical display) to the -400 (LED display) series aircraft. Their competency was assessed by instructors and training pilots and there was also a self assessment point where the flight crew had to confront their own ability to operate this new series of aircraft. When the aircraft developed a rough running engine on a B737, it was procedure on the -400 series to use the LED vibration dials to diagnose which engine was at fault. On the earlier series of B737 this gauge was notoriously unreliable and crews were advised not to rely on them for diagnosing engine faults. Instead a trial and error solution was used which involved easing the throttle back on either engine to see if the rough riding ceased. Unfortunately, in the G-OBME example, this technique was applied as the flight crew had not been trained to utilise the new instrument. This meant that when the healthy engine was throttled back and the aircraft levelled out, it temporarily cured the vibration. This misled the flight crew into believing that they had shut-down the correct engine.

The crew believed that the risk of their training letting them down was sufficiently minimal as to be acceptable. They further decided that the risk of making a longer diversion to East Midlands Airport (which was their maintenance base) was no less acceptable than making the diversion to Birmingham. In the event, the distance flown may have had no influence on the outcome, but the terrain on approach may have been significant in terms of the survivability of the accident. (The aircraft collided with a steep motorway embankment and broke up.) If the crew had held enough information about the incident (i.e. that the engine they were using would break up on approach) such that they would be able to choose which airport to divert to, then arguably, the accident would not have taken place. It is impossible to suggest that this risk taking decision was fallible even though the ramifications of the choice may have been less serious if they had selected to divert to Birmingham.

Finally, some of these decisions represent the sort of subtle risk-taking decisions which are made almost subconsciously on an hour to hour basis. In the context of being dissected after the event, they are often decisions that may seem very obvious, but this is rarely the case at the time these choices are made.

In the case of the announcement made by the Captain which informed the cabin that there had been 'some trouble with the right engine' when a number of passengers and cabin crew had witnessed torching or sparks from the left engine, the decision regarding whether to speak up and question the action was answered quietly within the confines of the individuals' minds. Confirmation bias could have persuaded the passengers that they actually heard *left* instead of *right*; Perhaps the engines are labelled facing backwards (in which case the left engine would be right) and so on? May be it was a slip of the tongue by the Captain - surely he knows how to fly the aircraft? No-one questioned the action except in their minds where they all concluded that the risk was acceptable. In hindsight, the decision was fallible, but entirely understandable.

Reasons for Variation in Risk Perception

Slovic, Fischhoff and Lichtenstein's first study (1980) of perceived risk concluded that both experts and laypeople '...differ systematically in their perceptions, ...particularly in regard to the probability and consequences of catastrophic accidents.' They add that 'Cognitive limitations, based on media coverage, misleading evidence, and the anxieties generated by the gambles life poses, cause uncertainty to be denied, risks to be misjudged, and judgements to be believed with unwarranted confidence'. In other words, the supply of poor information regarding risks has the expected consequence of fallible decision making.

The power of media coverage is echoed by Kone and Mullet (1994) who found in favour of a '...practically totally determinant effect of the media in risk perception'. In attempting to answer the question as to whether differences in risk taking behaviour solely a result of the information that has been provided, it should be asked whether, if supplied with similar levels of information, would two individuals make the same decision about risk acceptability?

Research conducted by Flynn, Slovic and Mertz (1994) to examine the influence of gender and race on the perception of environmental health risks took the unusual step of examining the characteristics of the 'risk perceivers'. Although some work had previously differentiated between the perceptions of men and women, the authors are only able to cite one study which look at racial differences. That study by Savage found that '...Blacks felt more threatened by Whites by each of four hazards: commercial aviation accidents, home fires, automobile accidents and stomach cancer.'(Flynn *et al.* 1994) Flynn *et al.*'s attempt to examine the effects of race on the perception of environmental health risks used a population of 1,512 Americans which were divided up as either Hispanic, White, Black, Asian or American Indian. Any study which delves into issues of race and ethnicity relies on self-definition and as such, the answers could not take account of what generation White or Black Americans might be.

The findings of the Flynn *et al.* study were that non-White males and females are much more similar in their perceptions of risk than are white males and females. As such, white males stood out, not least for consistently avoiding rating hazards as posing a *high* risk. As risk avoidance behaviour is largely based on perception, it may be the case that White males accept higher levels of risk or narrower safety margins.

A study of 38,829 male versus female pilot error accidents in 1986 found that not only were '...females significantly safer pilots as far as accident rates were concerned ...but they also kill themselves off at a significantly lower rate when they do have pilot-error accidents.' (Vail and Ekman, 1986). Whether this is a link to risk perception or a function of other factors such as the recruitment process or the motivations involved in working in a predominantly male dominated environment is open to speculation and therefore suggested as a subject for future research.

Kone and Mullet (1994) summarise attempts to examine perception of societal risk in different countries by highlighting differences between the USA, Soviet Union, France, Poland, Norway and Hungary. (See Mechitov and Rebrick, 1990; Teigen, Brun and Slovic, 1988; Englander, Farago, Slovic and Fischhoff, 1986; Karpowicz-Lazreg and Mullet, 1993; Goszczynska, Tyszka and Slovic, 1991.). In attempting to explain the significant differences the authors mention a discounted hypothesis which suggested that a deciding variable was the size of the country involved. They also offer the hypothesis that the active variable was the influence of the media. Responses from the former communist bloc countries (Soviet Union, Hungary and Poland) where accidents were rarely reported appeared to support the view that perceptions were related to media coverage. This was further supported by the authors' study of differences between Burkina Faso and France, the former being a Soviet satellite.

The power of the media in heightening a perception of risk may not be a bad thing in terms of aviation safety, regardless of how frustrating it may be for operators to see them appear to make *something from nothing*. In Australia, the media reaction to aircraft incidents and accidents is traditionally high profile. Whether this is a function of a general lack of competing news or a high expectation of safety in aviation is debatable, although the latter seems more plausible.

Following the Monarch and Seaview accidents and the incident involving an Ansett B747 at Sydney in 1994, a number of features appeared in newspapers and on current affairs television programs about the apparent demise of aviation safety in Australia. Articles which talked of 'The aviation scandal' (The Age, 1995) and which suggested that Australia's safety record had '...flown into turbulence' (Herald Sun, 1994) were commonplace. As the Sydney Morning Herald seemed to most aptly capture it 'By World standards, the crash landing of an Ansett B747 at Sydney Airport in October was not a disaster... but by Australian standards, where air safety within the big companies is not so much a matter of pride as a basic characteristic unquestioned by the travelling public, the crash was the worst incident in recent memory.' (Sydney Morning Herald, 1994)

The intensified media profile assured public consciousness of aviation safety matters remained high. As such, aviation safety has also remained high on the political agenda, a fact which precipitated the '...broad strategy to improve air safety (regulation) in Australia' which was announced by the then Minister for Transport, Laurie Brereton.

(HORSCOTCI, 1995) Indeed it was a high public and political conscious-
ness of an apparent decline in aviation safety standards that prompted the
House of Representatives Standing Committee on Transport,
Communications and Infrastructure to conduct its Inquiry into Aviation
Safety in 1995.

   If the media in Australia is guilty of heightening the public's perception of risk in aviation, then it seems likely that it is a contributory factor in lowering the societal acceptability of risk in this area. As a consequence, the level of resources which are allocated to counteracting perceived risk in aviation are likely to be higher. For example, when the
Australian Civil Aviation Authority existed as a Government Business
Enterprise between 1988 and 1995, it was tasked with recovering all of its
costs from its users. This involved commercial services within the CAA
cross-subsidising the safety regulatory function. When the Civil Aviation
Safety Authority was created in 1995 with the devolution of commercial
service provision to the new Airservices Australia, it became funded directly from the Government as a single-function safety regulatory authority.

   The major importance of risk perception is its power as a driver of
risk taking; a process which involves balancing perceived hazard and benefit against the risk countermeasures required to achieve safety.

   The Council for Science and Society (CSS, 1977) suggested that an
individual's acceptability of risk is governed by a number of considerations
including;

- Whether the risk is voluntary
- Whether the effect is immediate
- Whether there is an alternative
- Whether it is experienced occupationally
- Whether the consequences are reversible.

These may provide a model to examine why acceptable risk for Australian
commercial aviation may be *artificially* low, especially when compared to
other transport modes and why, as a consequence, safety countermeasures
provide an additional margin of safety. Figures 7.3 and 7.4 consider commercial aviation against road travel in Australia.

Figure 7.3   Risk acceptability for commercial air travel

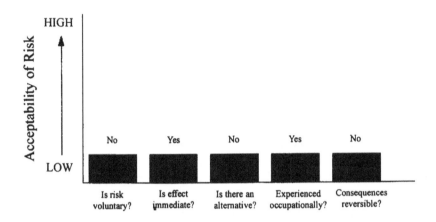

For the five aspects mentioned in the CSS model, acceptable risk is consistently low. In other words, safety countermeasures are demanded at a high level which in turn has an effect on the magnitude of safety margins.

Figure 7.4   Risk acceptability for private car travel

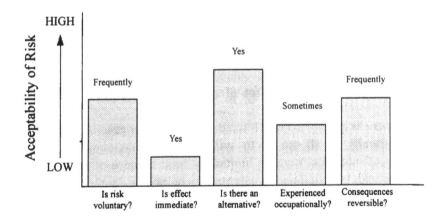

The risk profile for car travel is rather different with four out of five categories accepting a higher level of risk than civil aviation. The rationale for this profile is as follows;

Whether the risk is voluntary Long distance travel by air is judged to be a necessity by virtue of the distances involved in travelling both interstate within Australia and overseas. International passenger traffic is almost exclusively by air, although technically only 35% of interstate trips are by air (BIE, 1994). (The latter figure does not discriminate length of trip and is probably skewed by the large proportion of the population which lives near to the borders of Queensland, New South Wales, Australian Capital Territory and Victoria.) The only alternative for international travel is sea transport which takes several weeks to reach most destinations. As such international travel, whether it be for trade or for leisure purposes depends on aviation, and is therefore an involuntary risk.

Additionally, internationally air travellers are passive participants in the transportation, whereas a high proportion of car travellers are active participants (i.e. drivers) or able to directly influence the performance or behaviour of the driver. The apparent lack of direct control is one of the most commonly cited reasons for fear of air travel and is another reason why flying may be considered to be an involuntary risk. Individuals may perceive that if they were in control of a passenger aircraft, they would accept lower risk in certain situations, although in practice, their lack of technical knowledge and experience accounts for this difference.

Whether the effect is immediate In both car and commercial air travel, the risks tend to be immediate (with the exception of perhaps radiation or air pollution) from accidents. Such catastrophic risks tend to be more feared that the chronic threats of, say, heart disease, not least because a direct causal link between activity and outcome is easily apparent. As such the level of acceptable risk is lower for all transportation methods.

Whether there is an alternative Linked to the voluntariness of risk taking, there is virtually no alternative to aviation for international travel from Australia and interstate travel is limited by time because of the vast distances involved. Low numbers of alternatives reduce the level of acceptable risk as that activity becomes less voluntary. Car travel has a number of reasonable alternatives depending on the journey to be undertaken. These range from walking and cycling through to public transport. As such, the level of acceptable risk generally associated with car travel is higher.

Whether it is experienced occupationally The CSS (1977) found that risks which were experienced occupationally tended to be at a lower level of 'riskiness' than those experienced at other times. Aviation may be seen to have two groups of participants; namely the operators and the customers. Whilst the former are obviously experiencing occupational risk, the situation with customers is less clear. Business travellers may need to fly as part of their occupation whereas holiday travellers do not. Notwithstanding this, it may be argued that those who do experience occupational risk in aviation are in a strong position to demand a lower level of acceptable risk.

Historically, the high number of early accidents and incidents associated with aviation meant that the occupation of pilot was one with a high associated risk. As the vast majority of people involved in aviation were either pilots or at least from a flying background, the demands for improved reliability and safety were loud. This was particularly the case in Australia as mentioned earlier. This historical factor is possibly one of the reasons that risk acceptability in commercial aviation operation within Australia is relatively low.

Occupationally induced risk in car driving exists in some instances (such as travelling sales representatives), but in general, car travel is not associated with occupational risk. Even when a car is used to travel to and from work, it is not classified as an occupational risk and as such, the high proportion of leisure use accounts for a higher acceptance of risk.

Whether the consequences are reversible Some risks attract consequences which may be reversible over time. One example is car travel where injuries from non-fatal accidents are predominant and will often heal over time. In 1988 in Australia, only about half a percent of road accidents were fatal. (BTCE, 1992) Commercial airline accidents on the other hand, whilst rare events, tend to be fatal 50% of the time (Schiavo and Chartrand, 1997). Risk acceptability is set at a much lower level when consequences are believed to be irreversible.

Aviation is a safe mode of transport, especially when based upon accident statistics. However, there is a long-standing general perception that, in spite of such statistics, it is something to be feared. Certainly the potentially catastrophic nature of aircraft accidents and the potential for third-party victims adds to the fear. However, the heightened perception of risk may in fact have a significant positive effect on safety.

Where risk is overcompensated for, the by product is addition safety margins. Whilst some will argue that the cost of "over-safety" reduces expenditure elsewhere or leads to consumers choosing other transport modes with inferior safety records, there is no guarantee that the safety dollar works in this way. There is no evidence to support the view that reduced expenditure on aviation safety will be balanced by a commensurate increase in expenditure on safety elsewhere.

The market for risk is an imperfect one, especially within the transport industry. Far from being a negative for the aviation industry, the effect is actually one of demanding higher safety standards than other transport modes. It is an advantage that should not readily be given away, either through complacency or in the name of harmonisation.

8 Exploring the Human Environment

Training

'Flying an aeroplane is a learned skill. Learning is so involved in perception, thinking and memory that it is impossible to put it in a tight compartment. One psychologist has described learning as 'the process of being modified, more or less permanently, by what happens in the world around us, by what we do, and by what we observe'(Beaty, 1995). As such, it is difficult to evaluate the quality and effectiveness of training for an entire aviation system without very in-depth analysis. The background of employees in an RPT airline is varied and even the comparison of airline level training is difficult because of the numerous variable involved in the process of training and learning. Nevertheless, there are a number of issues of importance that relate to training which have had an impact upon aviation safety in Australia.

Recruitment

Would-be airline pilots within Australia have traditionally had few RPT carriers to aim for during their flight training. Although Ansett, Australian and Qantas have run cadetship schemes at some point in their history, they have not been the main source of pilots. Australian, for example, only had two cadet intakes, which were in the late 1960's, although Qantas did run a quite extensive cadetship program with 14 courses commencing between 1963 and 1970. (Stead, 1995)

As such, the main route to the airlines is through general aviation or the military. Baker (1988) estimated the traditional source of Australian Airlines pilots to be approximately 60% from the military, 35% from the armed forces and 5% from other airlines.

Selection was difficult, particularly as the recruitment process developed in line with increased competition. Whereas selection in Qantas before 1960 relied on flying experience, interview and medical examination, it has since developed to include a battery of psychological testing. Whereas candidates in the 1960's needed fulfil the following direct entry requirements:

- Commercial Pilot's Licence
- 500 hours minimum command experience (experience commensurate with age)
- Multi-engine time
- First class instrument rating.

These had extended by the 1980's to include senior commercial pilots licence theory subjects / rating and a twin-engine endorsement (with exception for military fighter pilots).

This change reflects not just an increased awareness of the attributes required in a good candidate, but also the increased availability of pilots to choose from. It is clear that one of the by-products of a large number of potential recruits is the ability of the employer to select a higher calibre of employee.

Recruitment is perhaps a more important concept than training as even the best training system needs quality candidates if it is to hope to do its job properly. 'Training is one thing, but the pilot must be trainable and have the potential of developing in due course into an aircraft commander.' (Davenport, 1988)

CASA stipulate a minimum requirement for pilot selection which is exceeded by Qantas at a cost of $230,000 per annum. (Lewis, 1997) A pilot will represent a long term investment which will require expensive training (thought to be in excess of $2 million to Captain in Qantas at present). The slow speed of progression through ranks may mean that without adequate selection tests, a *bad investment* may be difficult to spot. Davenport (1988) observes that the position of Second Officer in Qantas means that '...it may be three to four years before inherent flying deficiencies become apparent.' Recruitment is important to the Australian major carriers and their commitment is highlighted by their work on the development of customised psychological testing instruments. (see Alexander and Stead, 1993)

As most of the pilots who apply to join the major RPT carriers have paid their own way through training, there is little doubting their commitment. To accumulate sufficient hours and experience to be considered for employment, they will have had to spend a period of time flying charters, bank-runs and / or instructing, often in inhospitable terrain. 'The majority of airline pilots from Australia paid their own way through local flying schools, then went into the real world to sink or swim. This experience just may be one reason why that country has, at least to date, such an enviable safety record.' (Flight Internationàl, 1991).

Indeed, even Qantas cadets were sent out to *fly bush* upon graduation to gain much needed experience. In spite of anecdotal evidence regarding *good flying weather* and *flat terrain*, the hazards of General Aviation flying within Australia are significant. The level of accidents in this sector is high; averaging between 1.0 and 1.5 fatal accidents per 100,000 hours per year. (Commonwealth of Australia, 1997) Whilst aviation is a well respected career in Australia, it does not carry the same social status that it does in other cultures. Individuals become pilots because they want to, not because society ascribes a high level of respect to that profession. The slow route of experience which involves the *serving of time* in the harsh operating environment of General Aviation or the Armed Forces has a naturally selecting effect on those wishing to become airline pilots.

Evaluating Training Effectiveness

It is all but impossible to compare the training regimes of different airlines for a number of reasons. These include variations in;

- recruitment policy
- training style
- line flying environment
- technology (aircraft / simulators)
- organisational culture.

As such, it is impossible to say that training at *airline X* is only half as good as at *airline Y* because the courses are only half the length. It may be that recruitment at *airline X* specifies a higher standard and has a greater emphasis placed on on-the-job learning. For the latter to happen requires a committed organisational culture. In other words, it has to be possible for different members of the crew to help train each other. 'Not every airline will subscribe to the need to develop a training culture within the organisation' (Beaumont, 1997). Indeed some airlines would positively discourage training being done by anyone other than official training staff for fear of a lack of standardisation. The counter-argument within Qantas would be the strict adherence to standard operating procedures that is the cornerstone of their operation. The position of Second Officer, who is carried on all flight lasting over eight hours, fulfils a dual role as *cruise pilot* and as officer in training. Whilst the S/O may have many hours of flying experience (including a minimum of 500 hours in command), they will stay in the position for about 2-3 years watching and learning from the other line crew members. By the time they reach First Officer, they will have received an unusually high amount of experiential training.

Comparative evaluation tools are limited and regulatory licensing procedures for monitoring standards examine compliance and not any *additional quality*. Instruments such as the NASA/UT Line - LOS checklist (Helmreich, 1991; Hines and Helmreich, 1997) have only been used quite recently and tend to focus on a particular aspect of performance such as CRM skills. In applying such a tool, Helmreich (1991) was surprised to observe '...great variability among crews operating the same type of aircraft' even in the comparatively highly regulated USA. Although NASA/UT have conducted extensive testing using the L/LOS checklist, it does not provide a comparative tool that can be used easily. Observers need to be trained and standardised and fly with numerous crews on numerous sectors, which make it a very expensive process. The checklist has yet to be used in Australia, although Ansett is planning to do so.

There are no readily available primary markers currently available to support the view that training within the Australian RPT system is above average. However, a lack of accidents would seem to support a consistently high level of skill. Boeing observe that flightcrew error has been the primary cause in around 70% of accidents (Lautman and Gallimore, 1987) and whilst no training regime could currently claim to be able to prevent all of these factors, the disparity of accident rates between carriers suggest that more effective training is possible. 'Flight crew members are highly disciplined professionals and they respond to emergency situations in the manner in which they have been trained.'(Doss, 1992)

Australia's reputation for good RPT safety is based on a lack of accidents. A breakdown of accident statistics from around the world highlights flightcrew performance as the single greatest primary cause. As such, if failures to the Australian safety net were to occur, this is the area which is most likely to be deficient. The lack of accidents suggest a strength in the effectiveness of crew training and the training of support staff such as engineering and maintenance and air traffic control.

Level of Training

Terrell (1988) comments that one of the main elements of safety management within Qantas is the issue of training. As such Qantas has invested in training facilities which include;

- Advanced flight and ground simulators which expose crews to realistic emergency conditions too dangerous to practice in an aircraft.
- Emergency procedures training in which evacuation procedures are taught to pilots and cabin attendants.
- Maintenance training simulators for personnel involved in the line maintenance of aircraft.

Australian Airlines and Ansett Australia also had high-fidelity flight simulators based in Melbourne. Where equipment is only held by one airline, co-operative agreements exist to allow staff from the other airline to use it. Some of the subsidiary airlines have also now acquired flight simulators for their Dash-8 and Saab 340 aircraft.

The level of training is not just about the equipment which is available, but is a function of pre-employment experience, formal *ab initio* training and on the job training. It is supplemented by the use of CRM training, LOFT exercises and Emergency Procedures training.

Controlling the effectiveness of training is a very strict standard operating procedure culture, which has been particularly evident in Qantas. The effect of personality and leadership style is minimised by strictly following the book procedure and feeling able to speak up if this is not done. The Qantas or Ansett *way of operating* is made clear from the recruitment stage, through training and in normal line operations. 'Right from the beginning it is important to ensure the applicant understands the company's goals and company's culture... it is a serious business. Teamwork and crewmanship are taught in the first few days'. (Davenport, 1988) 'Individual and group behaviour of crewmembers forms the tone of cockpit operations...

The degree of self-discipline and procedural discipline affect the overall cockpit environment and in turn, determines the level of risk at which the flight crew operates.' (ICARUS, 1994)

A low turnover of staff within the major carriers, a function of good working conditions and the prized nature of jobs, means that pilots in all positions have a high average level of experience. *Career First Officers* are quite normal in Qantas, as are First Officers with more flying experience that Captains from other major carriers around the world.

Experience and Currency

As mentioned above, competition for employment with the major carriers is intense and as such, they are able to make selections based on the most compatible experience. This does not necessarily equate to employing new recruits with the highest hours, especially in Qantas where *ab initios* may have comparatively low numbers of hours. Indeed, secondary level carriers such as Eastern Australia and Kendell generally demand higher numbers of flying hours as new recruits will go straight into flying aircraft as First Officers. Qantas, on the other hand has the ability to train and standardise its new intake through the *apprenticeship* Second Officer position.

International Operations

International services out of Australia represent probably the longest average sector length of any airline in the world (Learmount, 1987). The advent of long range aircraft in the form of the Lockheed Constellation, B707, B747 and B767ER has been the result of demand from carriers such as Qantas. Indeed, long range variants of the Constellation (L1049G), B707 (-338C) and B747 (-B) were built at the request of Qantas (Gunn, 1988).

Early long-distance services were limited by aircraft range and services such as the *Kangaroo Route* to London via Asia required several stops. For example, when the B707-138 commenced the service between London and Sydney, it would make intermediary stops at Rome, Istanbul, Tehran, Delhi, Bangkok, Singapore, Jakarta and Darwin. Qantas' B747-400 series aircraft now have a range of 12,300 km. As such, this allows long-haul routes to be operated with zero or one stop. Long sectors such as Sydney to Los Angeles (12,054 km), Singapore to London (10,873 km) and

Melbourne to Johannesburg (10,312 km) mean that for long-haul crews, the opportunity to handle to the aircraft at take-off or landing is comparatively rare. Indeed, they find that the opportunity may only arise once or twice in a month.

The threat is a lack of experience, especially in terms of local conditions such as weather or complicated approach procedures. The counter-argument is that the situation is mitigated by the use of simulators and a heightened awareness of safety threats associated with recency. Additionally, the openness of communication which exists within the flight crew supplements the available experience of any individual crew member.

Nevertheless, this hazard should be acknowledged, especially as Ansett Australia expands its international operations and as longer range aircraft such as the B747-X and A380 become available and open up new route opportunities.

Domestic Operations

Although Australia's domestic carriers are faced with long distance *international-length* routes, the range of destinations for each aircraft type is relatively limited. For example, domestic B767-200s operated by Ansett Australia and Qantas may find themselves limited to the trunk ports of Perth, Adelaide, Melbourne, Sydney, Brisbane and Cairns. Even the more flexible types such as the B737 and A320 will find themselves limited to about 14-15 destinations Australia-wide.

Experience of operating into these key airports is high amongst domestic RPT carriers, assisted by a higher than average pilot utilisation rate than comparable airlines (see Table 8.1).

However, this is balanced out somewhat by the longer than average stage length operated in Australia (BIE, 1994) which dilutes the number of cycles which makes up the total hours flown. Notwithstanding this, the low number of destinations means that crews become very familiar with their routes. Whilst there is an ever present hazard of overconfidence, the approaches to major airports in Australia are, on the whole, straightforward and a high level of experience is the norm.

Table 8.1 Pilot utilisation in 1992 (Domestic)

Airline	Annual hours flown / pilot
Australian	352
Ansett	310
VASP	267
British Midland	263
Alaska	259
Aerolineas Argentina	186
Aviaco	183

Source: Data from BIE, 1994.

Both Qantas and Ansett pay particular attention to their Check and Training and Quality Assurance systems for ensuring standards are achieved and maintained. However, check pilot training is not required by the regulator, even though the airlines do conduct such training. Whilst check flights are the subject of regulation, CASA requires only two check flights per crew per annum (one simulator and one route check). Qantas exceed this by 100%; crewmembers are required to undertake three simulator checks per year at a significant cost of $12 million per year. (Lewis, 1997)

Continuous Fleet Monitoring

One of the difficulties of assessing training effectiveness is the so called *Hawthorn effect* whereby an individual under observation performs differently to normal. Simulator exercises and check flights are designed as assessment tools and with failure potentially meaning loss of licence and loss of earnings, the need to perform perfectly is heightened. Even *non-jeopardy* assessments such as CRM LOFT are liable to suffer from individuals performing *for the camera*. What is required to supplement these assessment methodologies is a way of monitoring performance and behaviours continuously.

One of the solutions is the use of Quick Access Recorders (QARs) which which allow Flight Data Recorder information to be analysed for trends from predetermined exceedance limits. The advantage is two-fold as the system is capable of tracking both recurrent deviations by individual crew members or recurrent deviations at a particular location or on a particular aircraft type. It is not a punitive system, rather a training tool which picks up occurrences that might be overlooked by the standard training and checking system. QARs are discussed further in chapter 10.

Joint Flight Crew Training

Training in the areas of aviation human factors and CRM has been developing steadily since the early 1980's, when a number of key accidents highlighted the human fallibility of the aviation system (Cooper, White & Lauber, 1980; Helmreich & Foushee, 1993). Statistics suggest that human error is found as primary cause in the overwhelming majority (70-80%) of aircraft accidents (Boeing, 1996; International Air Transport Association, 1993), and all aviation accidents can be found to exhibit some form of human error as causal factor. The emphasis in finding a training solution to this problem was originally focused on the flight deck, with early CRM courses known as cockpit resource management. The shortfall of such courses is apparent with the benefit of some years of hindsight. They only addressed part of the problem, that which related solely to crew members working within the cockpit. In doing so they ignored the valuable contribution to safety and efficiency to be made by the inclusion of other personnel, particularly cabin crew, as an integral part of the operating crew.

The exclusion of cabin crew from the core operational team is still widely practised today. While most airlines have now altered the nomenclature of their courses to reflect the fashionable transition to crew rather than cockpit resource management, for many companies only the name has changed, and CRM remains as a form of cockpit crew, rather than total crew, training (Hayward, 1995, 1997). This results in broad differences in the safety attitudes and knowledge of crew members (Merritt, 1993). However, in recent years, a greater understanding of the multiple causes which lie behind complex industrial accidents has led to a recognition that human factors knowledge and attitudes are important in all aspects of the aviation system.

The British Midland B737 accident at Kegworth (UK Air Accidents Investigation Branch, 1990) and that involving an Air Ontario F-28 on take-off at Dryden, Canada (Moshansky, 1992; 1995; Maurino, Reason, Johnston, & Lee, 1995) are classic case studies of the perils of ineffective use of all available resources, and have been used to justify the integration of crew Emergency Procedures training at some air carriers (Baker & Frost, 1993; Chidester & Vaughn, 1994; Hayward, 1993). The critical feature of each of these accidents is that the operating crew were acting as two crews - one running the cockpit and the other the cabin - rather than one. This resulted in failure on the part of the cabin crew to inform the cockpit crew of vital information which may have prevented these accidents. This failure is an outcome of inappropriate technical training (National Transportation Safety Board, 1992; Transportation Safety Board of Canada, 1995; Chute & Wiener, 1996), which is in turn contributed to by a lack of awareness of the type and level of knowledge held by cabin crew, and misunderstanding of the roles and responsibilities of crew members.

Australian Airlines introduced a two-day Annual Proficiency Check Course (APCC) in 1991, which ran until the airlines merger with Qantas (where it continued under a new title). Developed from the original Aircrew Team Management program which was introduced in 1985, the course brought together 16 flight attendants and four technical crew for two days. The effect was significant; 'Crew member cohesion and support developed where little previously existed, mostly out of ignorance generated by each group training in isolation.' (Baker & Frost, 1993)

The significance of this type of training is two-fold. Firstly the role of flight attendants as part of the operating crew is elevated such that inter-action no longer stops at the cockpit door. This is not simply about making the aircraft a happier place to work, but rather extending the knowledge, skills and attitudes of all of the aircraft's flight crew such that the safety net is widened. Even in the first two years of APCC, the trainers noted a difference in feedback from crews complaining about each other in the first year to relating more positive incidents of support (Baker & Frost, 1993). Although there are no comparative scientific appraisals of cabin crew knowledge, skills and attitudes currently available, the importance of the Australian Airlines (and subsequently Qantas and Ansett courses) is highlighted by a number of secondary indicators. These include the fact that the Australian course was one of the first in the world and the region is the first in the world to have a Cabin Safety Working Group.

The second significant issue is the fact that both organisational and national cultures have allowed this type of training to take place. The role of flight attendants is something which means different things to different airlines. In the Australian carriers, the primary role of flight attendants is safety. 'The basics that Qantas requires in the flight attendants must be safety.' (Jensen, 1988) Anecdotally, this has sometimes appeared to be at the expense of service quality. One expert witness suggested that '...as a passenger on Qantas, he would be lucky if he got a cup of coffee thrown at him, but he knew damn well that in an emergency, the flight attendants would be there to throw him down the evacuation slide'.

The Australian Bureau of Industry Economics examined the performance of the Australian aviation industry in 1994 and examined differences in costs between the Australian carriers and their international competitors. Table 8.2 lists salary levels for 1992 and shows the difference between pilots and cabin crew. Although difference between locations may be expected because of the cost of living, this does not account for the level of disparity between pilot and cabin crew salaries. It may however give an indication of the difference of role; a factor which may well be cultural.

Table 8.2 Average airline wages, 1992 (US$)

Airline	Pilots Salary (US $)	Cabin Crew Salary (US $)	Cabin Crew as % of Pilots Salary
All Nippon	161,223	52,798	32.75
Swissair	191,352	48,515	25.35
SAS	125,643	59,625	47.46
Northwest	117,954	26,978	22.87
Delta	113,645	31,569	27.78
United	117,981	34,145	28.94
American	99,254	23,439	23.62
Cathay Pacific	241,881	30,104	12.45
Qantas	95,464	32,514	35.74
Air Canada	90,979	35,691	37.39
Continental	65,143	20,470	31.42
British Airways	106,464	29,408	29.62
Singapore	96,990	15,877	16.37

Source: Adapted from Bureau of Industry Economics, 1994.

The Scandinavian carrier, SAS has the lowest disparity with cabin crew earning 47.46% of the pilots salary. These represent *feminine* cultures where the emphasis is on quality of life and care of the weak (Smith, Dugan and Trompenaars, 1994). The *Anglo* carriers in the US, UK, Canada and Australia all tend to be reasonably similar with cabin crew earning between 22.87% (Northwest) and 37.39% (Qantas) of the pilots salary. Finally, the Asian carriers have the greatest difference with Cathay cabin crew earning 12.45 and Singapore earning 16.37% of the pilots' salary. These cultures are ones where male and female roles are traditional and strongly enforced; pilots are male and flight attendants are female. The role of cabin crew also tends to be aimed particularly at customer service and as such personnel tend to be recruited to be *pretty young things* rather than as part of the safety team. The salaries reflect the difference in attitude and expectations towards the role of cabin crew. It is worth noting that behind SAS, Qantas cabin crew are paid the largest proportion of pilots' salaries, which correlates with the emphasis placed on their safety role.

Communications

The primary language of Australia is English, which is also the preferred language of International aviation. As previously mentioned, it was the opinion of a series of expert witnesses that one of the most important factors behind Australia's apparently good record was in the openness and unambiguity of communications. Cultural theory suggests that the low power-distance index rating for Australia would translate into ease of communication between ranks. The high level of individualism would also suggest that communication would tend to be direct, which may mean by-passing formal communication channels where necessary.

To test the directness of communications within Australia, pilots were asked

A senior manager introduces a new company rule you consider to be unsafe, which of the following statements best describes your actions?

1. I would simply ignore the new rule; its my life
2. I would complain about the rule to my colleagues
3. I would complain about the rule to my union representative
4. I would complain about the rule to my fleet manager
5. I would complain directly to the manager responsible for the rule.

Figure 8.1   Results for Qantas Airways

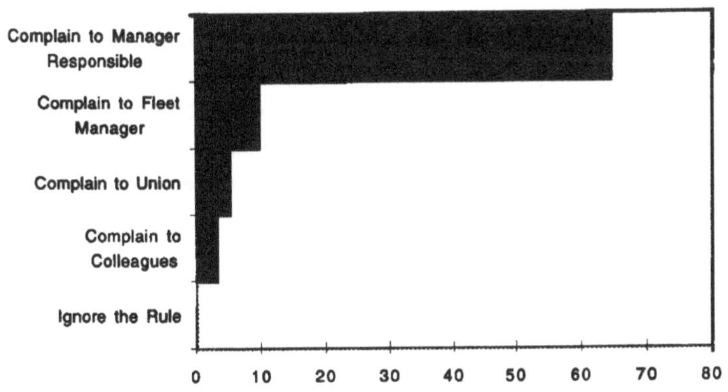

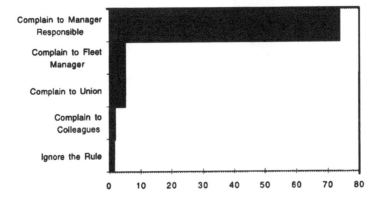

Figure 8.2   Results for Ansett Australia

The majority answer was quite clear. Qantas recorded 65% and Ansett Australia recorded 74% for the response that crew members would complain directly to the manager responsible. This is the answer predicted anecdotally before the survey. Max McGregor, Manager of Operational Safety for Ansett Australia suggested that in the event of the introduction of new operational rules, it was not uncommon for crew members to head straight for senior managers' offices if they disagreed with its philosophy.

The difference between the Qantas and Ansett response may be explained
by the geographical distribution of crews who may find themselves away
from home base for up to two weeks at a time and therefore unable to
approach senior management directly. It is also a much larger organisation
with a commensurately larger management structure.

The true significance of the results only becomes apparent once
compared to British pilots used as a comparison. Asked the same question
as the Australian carriers, the response was as follows;

Figure 8.3   Results for British airline pilots

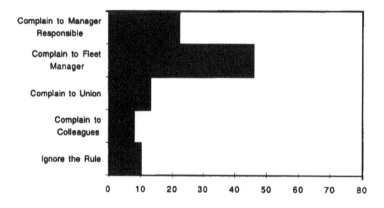

Only 22% of British crew members would go directly to the manager responsible, demonstrating a marked tendency towards adhering to
company hierarchy. This matches indications from existing cultural
research although the magnitude of the disparity is greater than was expected. It is frustrating that the research did not have the resources to encompass
comparative research groups from cultures which appear at the extremes of
Hofstede's cultural grids. This does, however, lay the groundwork for future
research and it is hoped that these results provide a good basis for that work.

The majority of British pilots (46%) would go to their fleet manager whereas only 10% of Qantas Crew and 5% of Ansett Australia crews
would claim to do this. Another interesting indicator is the number of crew
members who would action their grievance through the union. Only 5% of
Australian crew members would utilise this channel in the first instance

compared to 13% of British Crews. This points to the loss of faith suffered by the Australian pilots in their unions following the 1989 pilots dispute. (The strike led to a large number of domestic pilots losing their jobs.)

The British also seem more inclined to complain to their colleagues (8%) than the Australians (Qantas; 3%, Ansett; 2%). This would seem to fit with the cultural stereotype of 'whingeing poms' or rather of the slightly lower level of individualism that Hofstede's grid assigned to the British over the Australians.

What is of concern is the high number of British airline pilots who claim that they would ignore the rule; a staggering 10%. This is in contrast to 0% from Qantas and 1% from Ansett. This suggests that the Australians are not willing to accept dangerous operational procedures without doing something about it. They will not ignore rules, which is to be expected in the light of comments regarding the strict adherence to standard operation procedures.

A further variation of the question was included for the military transport crews to represent the different operational environment;

A senior officer introduces a new operating rule you consider to be unsafe, which of the following statements best describes your actions?

1. I would simply ignore the new rule; its my life
2. I would complain about the rule to my colleagues
3. I would complain about the rule to my Wing Commander
4. I would complain directly to the officer responsible for the rule
5. I would obey the rule as it is my job to obey the rules.

Military transport crews were expected to return a different set of answers and the question was constructed in such a way as to take account of this. The majority of pilots, again would go directly to the officer responsible for the new operational rule. Such direct communication is supported for flight safety matters and not considered to be insubordinate. The fact that 37% of the transport (generally non-combat) crew members would be prepared to obey rules they considered to be unsafe marks the contrast between military and civilian operations. Nevertheless, the result was greeted with considerable surprise by the collection of Air Commodores and Acting Chief of Air Force Staff who were shown it in a research briefing.

Figure 8.4   Results for RAAF transport pilots

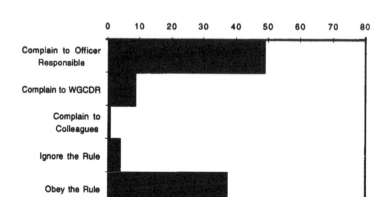

This represents an organisational culture trait within the military that is not unexpected, whereby junior rankings are expected to follow orders. However, the distinction between such compliance and the reaction to operational procedures (rather than tactical or strategic commands) is significant. Flight safety matters within the RAAF and Australian Defence Force (ADF) in general have received growing attention in recent years, not least in response to the loss of a B707 in a training accident in 1991 (DFS, 1994) and the Army Blackhawk helicopter disaster in 1996. In the former accident, the instructor attempted to simulate an assymetric two engine failure in flight, leading to the subsequent loss of the aircraft and crew.

Crew Resource Management Training

Crew Resource Management (CRM) first developed in the late 1970's as cockpit resource management through programmes run by United Airlines and KLM Royal Dutch Airlines. The original objective was to improve management of the cockpit, especially following the 1977 Tenerife Air Disaster where the KLM Chief Training Pilot had made the decision to take off in fog without air traffic clearance. Although the CVR tape revealed some concern from the subordinate crew within the KLM flight deck, there was no attempt to get the Captain to abort take-off and the ensuing crash claimed the lives of 583 souls. Original courses were based on business management training programmes (Härtel and Härtel, 1995) and have become ever more aviation specific over the last twenty years. The original courses also focussed almost exclusively on resources within the cockpit (Flight Safety Foundation, 1995) whereas modern courses aim to include '...extra-cockpit resources such as flight attendants and maintenance, air traffic control and dispatch personnel'. The latter development is now recognised in the working definitions of CRM.

Helmreich, one of the most respected experts in CRM, defines it as 'the effective coordination and utilisation of all available resources in the service of flight. These resources are both inside and outside the aircraft and are both material and human, including especially the knowledge, judgment and decision-making skills of all crewmembers' (Helmreich, 1987). In explaining the components of training required to achieve successful resource management, Diehl (1991) suggests, 'Such programmes stress the importance of proper communications, division of responsibilities, leadership and teamwork'. The author contends that the synthesis of CRM and ADM (Aeronautical Decision Making) create an agenda of five key areas for instruction, namely;

- Attention Management
- Crew Management and Communications
- Stress Management
- Attitude Management
- Risk Management.

However, one of the principal difficulties that has yet to be overcome in CRM training is how to assess the effectiveness of different courses or components of courses. Although the Flight Safety Foundation (1995) suggest that '...the importance of CRM has been demonstrated repeatedly in the performance (both positive and negative) of flightcrews during accidents', there are actually very few metrics available. (Conclusion of Royal Aeronautical Society Human Factors Group, 1996.) There are a number of anecdotal examples of aircraft that have been saved through the propitious use of CRM skills which are worth noting. For example, probably the most cited case involved a United Airlines DC10 which crash landed at Sioux City, Iowa with survivors, following what would previously have been deemed to be an unsurvivable loss of all hydraulic control lines.

However, most of these stories lack the sort of comparative data that may be required by the strict rules of empirical research. For example, the loss of B737 G-OBME at Kegworth, England in 1990 is often held up as an example of where CRM could have saved the aircraft. Whilst in theory this is probably true, it is impossible to make such a simplistic statement as 'this accident would not have occurred if the crew had received CRM training'. The fact that crew members have been through a particular training course is no guarantee that they will take on board and actively use those skills.

Returning to the example of the United Airlines DC10 crash at Sioux City; while the Captain of the aircraft now tours the world extolling the virtues of CRM, there were several factors involved in that accident that make it an extraordinary example. Firstly, the three man crew were augmented by an off-duty Check and Training Captain who administered the crucial thrust control and secondly, the performance of the crew was proved to be beyond any expectations. Following the accident, a number of other crews (all of which came from the same training background) were faced with a similar sequence of events in the flight simulator and none of them were able to land (or crash-land) the aircraft as successfully as the original crew. Whether this was a function of the individuals involved or the effect of adrenaline etc. on the actual day is beyond the realms of measurability.

Notwithstanding this, the general feeling is that CRM training is of very great value to the aviation industry. This is reflected in the number of programs which are now in place with major airlines and the move by certain regulatory authorities including the UK CAA to make the provision of such training mandatory. The 1997 Civil Aviation Safety Authority Regulatory Review Process looked set to mandate some form of CRM training in all airlines involved in commercial flying, although by 2000, this was yet to materialise. Prior to this, both Ansett and Qantas have established CRM training programs and more recently, regional carriers including Eastern, Kendell and Hazelton Airlines have been involved in setting up their own programs. It is now becoming standard practice to supplement CRM training for technical crew with courses that include cabin crew. This is sometimes done as part of recurrent emergency procedures training or as specific courses e.g. Ansett's Captains and Pursers CRM course. More recently, moves have been afoot to introduce CRM type courses for Air Traffic Controllers (Airservices Australia), Engineering and Maintenance Crews (Qantas) and Ramp Staff. (Qantas and Ansett)

Some of the problems that have been encountered when trying to translate the principles of CRM from its original host anglo culture indicate that certain aspects may need revision or change.

For example, anecdotal evidence from trainers working in certain South American countries indicates that CRM training has empowered First Officers to such a level that followership is becoming a problem and all of the Captain's actions are being questioned.

However, an important question is how dependent upon the organisational culture of the host airline or indeed national culture of the host nation, the success of CRM training actually is? To answer this, it is necessary to look at the fundamentals of what such training is trying to achieve. In simple terms, that is a shift in attitude away from working as a set of individuals to performing as a team. For example, Ansett's Ramp Resource Management course teaches its staff that '...it is better to be part of a champion team than to be a team of champions'.

Does this shift in emphasis from individual performance to group performance go against the strong cultural dimensions of individualism or power-distance? This is a question which is difficult to answer without significant multi-cultural investigation. However, it is worth posing some of the questions here, partly to encourage further research and also to suggest explanations for the variability in acceptance of CRM.

Individualism This dimension may explain more about the need for CRM training than just its variability of success. Highly individualistic nations ascribe success to the performance of the individual rather than of the group. Even in areas where group performance is supposed to be very important such as in sport teams, there is a tendency to elevate key individuals. For example, the level of hero worshipping for individuals such as Eric Cantona (France, Soccer) or Michael Jordan (USA, Basketball) highlights this process. Although civil aviation has long since required multi-crew on large aircraft, it still bases flight training on individual performance. This is not necessarily a bad thing, but has implications for the ease of introduction of training such as CRM.

Wolfe's (1979) legendary work, *The Right Stuff*, which charted the progress of USAF Test Pilots up to the implementation of the Apollo Space Program, has been largely responsible for highlighting an industry stereotype known as 'the right stuff'. This description captures the 'gung-ho' macho mindset of a group of pilots who believed that 'the right stuff' was either something they naturally had, or didn't. Those who were killed in flight, for whatever reason, were deemed to have the wrong stuff. Whether this was a strategy for dealing with grief or hiding fear is a matter for psychology to tackle, but it has become the epitome of what CRM is trying to cure. Recruitment still attracts this model of pilot, not least because the recruitment process has not evolved to select out such traits and interviews may be conducted by a generation of pilots which remain cynical to the benefits of CRM.

Does individualism affect the ability of pilots to work as a team and thereby embrace the principles of CRM? If that question could be answered then it may give some light as to how much CRM is needed in different airlines or countries. It is possible that some of the cynicism towards its introduction is based upon already applying the principles in normal operations.

On a 1995 observation flight between Singapore and London, the Qantas Captain explained that he thought '...CRM is a load of crap'. As the observer remained on the flight deck for most of the sector, he looked for signs of good or bad CRM. The Captain was not aware that he was under such examination and neither was the exercise conducted using an evaluation tool (such as NASA/UT's Line LOS Checklist). However, it was clear that in normal operations, the Captain was embracing the principles of CRM admirably. This included open and friendly communications with a new, female Second Officer who was part of the four-strong crew. Whilst it is impossible to suggest that this behaviour would necessarily be replicated under a high pressure, safety critical situation, it did highlight the possibility that the principles of CRM were being used as standard *airmanship*. This represents one of the lesser referred to side-effects of CRM; that is the risk of alienating the believers. Helmreich (1992) refers to *boomerangs* - non-believing pilots who do not respond to CRM training and as such, perhaps an awareness of *positive boomerangs* - believing pilots who do not respond to CRM training, is required?

Power-Distance CRM attempts to improve team performance through the opening of communication channels. This is achieved through the training of "appropriate assertion" and other related communication techniques. Crew members are made aware of the technicalities of style from the role of body language and written word, to the use of empowered phraseology such as 'Captain, you must listen!' which is used by Qantas as a final level of assertive upward communication. Cultural research has demonstrated that the process of communication between ranks is not just dependent on the structure of formal communication systems, but also the cultural context it operates within. The concept of Hofstede's Power-Distance index is based on national culture difference rather than organisational and yet indicates the relative ease of which communications can occur.

Australia records a low Power-Distance ratio such that questioning of authority is neither disrespectful nor causes loss of face. This is in contrast to countries such as Japan or China where respect for seniors can literally become a situation of life or death. In an incident where a Captain decided to kill himself by flying his aircraft into Tokyo Bay, the First Officer is alleged to have politely requested the Captain not to do so as crashing the plane may damage the company's aircraft or harm the passengers in First Class. At no point did the First Officer attempt to take over the controls or even raise his voice and the aircraft subsequently crashed.

An important aspect of organisational culture within Qantas was the empowerment of junior crew members through standard operational procedures. The most frequently cited example is often considered to be reflective of national cultural traits, although deeper investigation suggests this is only part of the explanation: Qantas has always carried Second Officers (S/O's) on the flight decks of its international jet aircraft (B707, B747, B767) to fulfil several roles. These included training and the accumulation of flying hours and as relief pilot in the cruise phase on long distance routes. The Second Officer is a licenced air transport pilot, but will only fly the aircraft in cruise. A test of the power gradient on the flight deck is the ease of which junior crewmembers can speak up if they believe their senior crew members have made a mistake. A number of expert witnesses highlighted the fact that junior crew members on Australian flight decks are able to speak up in this way without fear of reprisal or causing loss of face to the senior crew member. (Lewis, 1995; Green, 1994)

The reasons for this appear to be two-fold. Namely that cross checking of senior crew members by the second officer was proceduralised within standard operational procedures as being part of their role and that the national culture context was relaxed enough to allow this rule to work. Although Terrell (1995) suggests that the rule was introduced partly '...to give Second Officers something to do...' in long, uneventful cruise phases, there was an early recognition within Australia of the benefits of a skill which CRM has tried to encompass; namely that of cross checking of crew members actions. Ashford (1994) highlighted the fact that in 54% of 118 large-jet accidents, there was inadequate cross-checking by other crew.

Due to the mix of domestic and international flights that Qantas now operate, especially using the B767, the role of the Second Officer no longer explicitly dictates that the cross-checking of flying pilots. However, it is a practice that is firmly entrenched in the company's culture and done as a matter of routine. This is a good example of where an aspect of company culture has been established over time and remains in place even after written procedure has been changed. The procedure of cross checking did not disappear from print because it was thought to be obsolete or incorrect, but because the activity was now inherent to the way that a Qantas Flight Deck was operated. It is reinforced by the training process and attitudes of senior crew members.

Organisational / Industry Structure

A significant factor which both affects, and is affected by, communications is the structure of organisations, and indeed, the industry that a particular organisation may find itself operating within. As there are very few unknown accident types, only variations on original themes, it seems fair to assume that repetitions occur because individuals possess incomplete or inaccurate information. There are few unsolved accidents so why does the industry keep making the same mistakes?

Short term, selective memory is not unique to the aviation industry. Beaty (1995) highlights the '...need to look at mistakes in a much wider context instead of in constrained and separated specialist enclaves with little communication between. ...Individual expertise in the different environments of air, sea, road, rail and ground are of course essential, but a connection between them and a cross-fertilisation of information on the pivot of a corporate understanding of human factors needs to be established.' This is because not only are similar, systemic and organisational failures occurring across different modes, but also there are lessons that have already been learned elsewhere which do not need recreating. Reason's (1992) examination of organisational accidents derived from a general psychology background to see the same sort of latent defects and active failures occurring in accidents as diverse as the Bhopal Union Carbide Chemical Works, The Challenger Space Shuttle and the Kings Cross Underground Fire.

The issues here are twofold; namely the effect of national culture on hierarchy and communications and the effect of organisational structure (which in turn may be directly influenced by national culture). The former has been discussed in considerable detail in previous sections. The safety (efficiency) message in the case of organisational communications is a simple one; that the corporate structure must be set up to facilitate and encourage open communication. FAA Assistant Administrator for System Safety, Christopher Hart (Male, 1997) '...believes that the only way of further reducing airline accident rates is the sharing of safety information.' In other words, the expertise is generally in place, it is just not communicated to the right people at the right time.

However, even if communication channels appear to be in place, the integrity of data needs to be assured. Bent (1997) suggests that 'Without data, you are just another person with an opinion' and yet all too often 'fact' is apparently borne out of enough people saying the same thing (Braithwaite, 1997). While a commitment to the goal of 'safety first' may be common, the level of commitment varies between individuals and organisations. Accidents only occur when decision makers have perceived risk taking decisions to be acceptable in terms of safety. 'People respond to the hazards they perceive'(Slovic, Fischhoff and Lichtenstein, 1980) and consequently if these '...perceptions are faulty, efforts at public and environmental protection are likely to be misdirected.'

Efficient and effective communication is the key to avoiding misperception. Efficiency is assured through establishing channels of communication in the organisational structure and the way in which communication is encouraged. Qantas' (1995) Crew Resource Management course suggests the following list of effective communication rules, some of which translate directly to the question of organisational structure;

a)   Be descriptive, specific and non-judgmental in reply   This reflects the need for any communication to contain the correct information and is a function of both formal communication systems (such in incident reporting forms) and the organisation culture in the form of precedence. Judgmentalism may also be a function of this and can be influenced to a degree by traits of national culture.

b)   Choose an appropriate time and place   Formal communications structures such as forums or meetings may provide an appropriate time and place, but need support to be arranged and should be backed up with other informal channels. For example, communication between Ansett and Qantas Safety Departments is continuous regardless of when official forums such as the Australasian Flight Safety Council may meet. This process can be hampered by geography - Australian Airlines and Ansett were based in Melbourne and Qantas in Sydney although advances in telecommunications and the availability of travel have helped this. More of a problem exists when dealing with international concerns such as airframe or engine manufacturers and other operators. This has always been the case in Australia which from the start of European settlement depended on Europe and the USA for a technological lead. However, the extreme delay in communication at this early stage forced the need for self-dependency which in turn had an effect on how the Australian industry dealt with problems. Expertise within the Australian system was developed to cope with both the spatial separation and time-zone differences which exist.

c)   Tailor the length and content of communications to suit the occasion
One of the main problems for effective communication is making sure the important aspects of the message get through. In transport operations, this becomes even more important when information may be time critical. For example a go-around command or collision avoidance call may need an instantaneous reaction and the aviation industry has constructed standard phraseology to allow this to happen. As English is the common language of aviation, there is an extra margin of safety afforded to individuals where it is their first language. For example in the 1977 Los Rodeos accident where two KLM and Pan American B747's collided on the runway, the First Officer uttered a non-sensical reply to the Air Traffic Controller when asked about his status. The Controller had instructed the KLM B747 to hold at the departure threshold, but the KLM Captain elected to take-off without clearance. The phrase 'We are now at takeoff' was intended to mean 'We are now taking off' whereas the controller interpreted it as 'We are now ready for the takeoff' (Cushing, 1994).

Airlines train both standard phraseology for use with technical communications (such as to the air traffic control tower) and in other situations. Qantas ensure that junior crewmembers feel empowered to speak up in a conflict situation through aspects of its CRM training course. Appropriate assertion is taught as a skill and supplemented by a 'last resort' power phrases whereby any crew-member can guarantee the attention of the aircraft captain. Such a proceduralised approach is the exception rather than the norm and is reflective of the need to support practices with official procedure.

d) Create an environment which encourages people to communicate Even where a formal structure is put in place to facilitate effective communication, organisational practices can mean that the objective is not met. For example, mandatory incident reporting can suffer from lack of reporting when individuals perceive that the process does not offer them anything in return (i.e. lack of feedback). Encouragement may be in the form of financial support e.g. to attend conferences, working groups or forums or may be through training courses such as CRM or Operational Teamskills (Ansett's version of company-wide CRM type training) where the need and methodology for inter- and intra-group communications are highlighted. Openness in communication is a strength in Australia and within the Australian aviation industry which exists at a cultural level. As such it may the environment which encourages communication may be invisible, except under close comparison.

e) Avoid ambiguity Ambiguity can be the result of incomplete knowledge or understanding or may be the result of poorly phrased or translated messages. It can be avoided through attention to detail in *ab initio* and recurrent training, a developed systems knowledge and through structured communications protocol. The latter may be in terms of form design for written systems or standard phraseology for verbal communications. Whereas English speakers have an advantage, it is also an ambiguous language. 'Ambiguity is an ever present source of potential (air-ground) misunderstandings.' (Cushing, 1994)

Simplified standard operational procedures are designed to reduce this risk although experience with the introduction of SIDs (Standard Instrument Departures) and STARs (Standard Arrival Routes) demonstrated that unless standard procedures are exactly that, they are prone to misinterpretation. Australia is certainly not immune to risk from deficient communication as recent accidents have shown. It is important for operators to recognise the strengths of the existing system and fortify them if safety is to be maintained.

As mentioned above, organisational structure plays an important role in facilitating communication and therefore in assuring operational effectiveness. Over and above this is the role played by the industry structure which may be influenced by a number of geographic, economic and social factors.

Industry Structure

Australia represents a small population divided over a large surface area which in turn has led to one of the lowest areas of population density in the world. The entire population of Australia is roughly equivalent to the number of people contained within the London M25 Orbital Motorway on a week day, spread over a land mass roughly equivalent to continental USA. As such it is relatively unique and has an aviation industry which is commensurately different. This is not simply a feature of the historical influences of geography and demographics, but also the influences of political phenomena such as the two airline policy and economic forces. To what degree the structure of the aviation industry has played a role in safety is difficult to quantify, yet is worth consideration to better understand the mechanisms of systemic safety.

Communication between the major carriers on matters of safety has always been good. Although Australian Airlines (and latterly Qantas) and Ansett were always in competition, their safety professionals worked together, not least because of their geographical isolation from other similar carriers and aircraft manufacturers. Co-operation existed on a formal level through bodies such as the Australasian Flight Safety Council and the Australian Society of Air Safety Investigators and at an informal level between individuals from each carrier. Ansett Manager, Flight Safety; Australian Airlines Manager, Flight Safety and Qantas, Head of Safety and Environment would talk to each other all but every day, even at times when their relative commercial departments were locking their corporate antlers.

When an Ansett BAe 146 suffered a four engine roll-back in unusually warm air at high altitude near Meekatherra, WA, the Ansett Manager, Flight Safety not only contacted the Regulatory Authority, but also contacted his opposite number at Qantas. Regional services operated on behalf of Qantas by Airlink utilised BAe 146 aircraft and both Ansett and Qantas voluntarily imposed a maximum operating ceiling on the aircraft type well before the CAA made any such recommendation. This is but one example of similar positive action well before the regulator made a move.

There is a friendly rivalry in existence between the two safety departments which allows them to co-operate rather than conspire. When the Ansett B747-300 VH-INH landed at Sydney Airport with its nose wheel retracted, in full view of the Qantas Safety Department, there was a genuine rush to communicate the news to the Ansett Manager, Flight Safety. Once it had been established that no-one had been injured, the communications became perfect examples of Australian 'larrakin irreverence'.

Unfortunately, communication within and from the CAA was severely damaged by the reforms of the late 1980's and early 1990's. Within the organisation, the reaction was one of '...uncertainty, fear, resentment and antagonism towards senior management and between CAA staff.' (HORSCOTCI, 1995) These were hardly the best conditions for good communications at a time when both the organisation and the aviation industry were experiencing severe change. Westrum (1995) writes that 'aviation organisations require information flow as much as aircraft require fuel.' Infighting within departments about management positions was rife, particularly in the Safety Regulation and Standards division and between air traffic controllers and flight service officers. One of its former chief airworthiness surveyors, Laurie Foley commented that '...the culture of the organisation is the basic problem. ...Until you change the basic culture of the organisation, you won't fix the problem. The recipe might have changed, but the ingredients have stayed the same.' (The Australian, 1995)

Direct communications from the Board Chairman, Dick Smith were extreme examples of the Australian cultural trait. A number of CAA employees, who wished to be deidentified, cited examples of receiving phonecalls from the Chairman telling them to do something. When they tried to explain why certain things were technically impossible, their protests fell on deaf ears. Arthur Jeeves, an airworthiness inspector with the CAA told the HORSCOTCI that staff who did not share Smith's vision were told that they should leave. 'His dictum was both fundamentalist and amateurish but he had the support of the minister of the day and the dye was cast. Those who dissented were virtually driven out.' (Herald Sun, 1995)

Poor communications between the CAA and industry were highlighted in the Seaview and Monarch accidents as findings in the BASI investigations. In the case of the Monarch accident, the report stated that 'Latent organisational failures identified within the CAA included ineffective communications between the local CAA District Office and Monarch.' (BASI, 1994). In the Seaview accident, the report observed that in relation to the surveillance activities of the CAA, 'deficiencies were treated in isolation from those which had previously been identified' (BASI, 1996). One of the reasons for this was likely to have been the fact that 'CAA district offices displayed significant differences in their respective operating practices'. These examples highlight the incoherence of an organisation in poor health and are perfect examples of the precursors or latent defects highlighted in the Reason model (Reason, 1990).

In Table 8.3, Hayward (1997b) illustrates the three distinctive patterns of aviation organisations as suggested by Westrum (1993, 1995) (pathological, bureaucratic and generative) using a chart of basic organisational communication styles.

Table 8.3 Basic organisational communication styles

Pathological	Bureaucratic	Generative
Information is personal power	Information is routine	Information is seen as a key resource
Responsibility is shirked	Responsibility is compartmented	Responsibility is shared
Messengers are shot	Messengers are listened to if they arrive	Messengers are trained
Bridging is discouraged	Bridging is tolerated	Bridging is rewarded
Failure is punished or covered up	Organisation is just and fair	Failure leads to inquiry / learning
New ideas are actively crushed	New ideas present problems	New ideas are welcomed

Source: Adapted from Hayward, 1997; Westrum 1993, 1995.

Activities within the troubled CAA represented a pathological style; information became personal power in the course of infighting and power struggles within the SR&S and air traffic services divisions. Messengers were metaphorically shot if they disagreed with the high level changes; directives and failure was covered up either through naïvety, overwork or the desire not to be the messenger of bad tidings.

On the other hand, communications within the major airlines tended more towards a bureaucratic or generative system. Information as a key resource is embraced through the principles of CRM, which in turn has been shown to be as much a function of organisational culture as it is a separate safety initiative.

A good example of how failure leads to inquiry / learning is provided by the Qantas Engineering and Maintenance Department. Human Factors incidents are examined through a "no-blame" inquiry. Following an incident, personnel who may have been involved are invited along to a hearing which is made up of management representatives from Engineering and Maintenance and the Safety Departments, and the trade union. The session is designed solely to establish what went wrong so that reoccurrence can be prevented through retraining or change of procedure. It is a process that is entirely separate from any investigation that may be conducted by BASI and is predominantly used for events which would not warrant a formal investigation by the Bureau. The system works because it is non-punitive in nature and allows all concerned to express ways to prevent future incidents. In this way, new ideas can be communicated between shop floor and management.

More recently, BASI has introduced a proactive safety program for regional airlines called INDICATE (Identifying Needed Defences in the Civil Aviation Transport Environment) which allows airlines to '...critically evaluate and continually improve the strength of their safety system' (BASI, 1996b). This is accomplished by the use of focus groups and has been successfully implemented at Kendell Airlines (a wholly owned subsidiary of Ansett Australia) and is being introduced at a number of other regional carriers. Indeed, the INDICATE program has now been endorsed by IATA as a safety management program. The importance of this issue is that it highlights organisational traits that needed to be present for airlines to agree to participate in the first place. Generative organisations welcome new ideas, information is seen as a key resource and failure leads to inquiry and learning (Westrum, 1995). The fact that Kendell and therefore its parent company, Ansett was prepared to participate (and actively support) the INDICATE program is evidence of a Generative organisational communication style.

9   The Lucky Country...

Political Influence

Aviation's importance as a transport mode within Australia is second only to road transport (BIE, 1994) and boasts an output which is as large as both rail and water transport combined. In 1992 the air transport industry employed 36,000 people with an additional 7,800 people employed in industries supplying services to the industry. (e.g airports and navigation)

It is an industry which is primarily involved in the movement of passengers, but with an important role in moving high value, time sensitive freight; a sector which has been growing at approximately 20% per annum. In terms of domestic travel, air represents only 35% of all interstate trips (BIE, 1994), yet it accounts for just under 100% of international trips. (BTR, 1993; ABS, 1993)

The monopolistic position of aviation as the mode of international passenger travel assures a high political profile for aviation. This is the case in most developed industrial nations, especially in terms of medium and long range trips, but particularly so for Australia because of its size and distance from other countries. International trade is highly dependent on air travel because of this reason, as is domestic interstate business which is '...a relatively intensive user of domestic passenger airline services.' (BIE, 1994)

This chapter highlights some of the political issues behind aviation safety in Australia whilst attempting to avoid becoming distracted by micro-political issues which may be linked indirectly with safety. As such, the priority of aviation in Australian political life is considered, as is the role of the regulator; a subject of vociferous debate within Australia over the last few years.

Priorities for Aviation

Aviation has become depended upon within Australia to a level which is unusual, even amongst developed nations. Not only does the *tyranny of dis-*

tance dictate a strong trade need, but it has also made the provision of services such as the Royal Flying Doctors Service a fundamental part of the social structure.

The high level of political interest can be noted through regulatory issues such as the *Two Airline Policy* and the strong intervention surrounding the Monarch and Seaview crashes.

However, it is first worth noting the position of air transport as a modal choice within Australia. As it is Business travellers who account for the largest proportion of RPT-level airline income and therefore also possess a great deal of political weight, it is worth considering their views regarding aviation's attributes within Australia.

In 1993, the Bureau of Industry Economics (BIE, 1994) surveyed 84 members of the Business Council of Australia regarding the most important aspects of aviation services. The results are summarised in Table 9.1.

Table 9.1   **Business assessment of attributes of air passenger services, 1993** *(average scores out of 10, ascending scale)*

Aspects of service quality	*Importance*	*Australia*	*Importance*	*Australia*
Safety	8.3	8.2	8.1	8.4
Timeliness on arrival	8.2	6.9	8.3	5.9
Timeliness on departure	7.8	6.9	7.9	6.3
Flight time	7.4	7.1	7.5	7.0
Comfort	7.4	7.6	7.0	7.5
Clearance time: baggage	7.3	5.4	-	-
Clearance time: check in	7.2	6.1	9.0	9.0
Fares	7.2	5.0	7.5	4.5
Frequency of service	7.1	6.6	7.4	7.1
Clearance time: immigration	7.1	6.0	n/a	n/a
Clearance time: customs	6.9	6.0	n/a	n/a
Airport services and facilities	6.5	6.2	6.4	6.8
Courtesy of staff	6.4	6.7	6.4	6.6
Liability coverage	5.7	5.9	5.6	6.0
Claims procedures	5.2	5.6	5.1	5.6
Average		6.4		6.7

Source: Adapted from Bureau of Industry Economics, 1994.

Safety was rated very highly in both importance and Australia's performance. For international travel it was considered to be the most important factor and the ranking indicates a high level of satisfaction. For domestic travel, *safety* was considered to be the third most important factor behind *clearance time: check in* and *timeliness of arrival*. Australia's performance for *safety* was rated above its importance ranking, second only to *clearance time: check in*.

There is an apparent disparity between the importance ranking of safety for domestic and international travel. While both score very highly, time issues become more important for domestic travel. This may hint towards assumptions regarding safety within Australia which are higher than for certain overseas routes or carriers.

Role of Regulator

The Australian Government took control of civil aviation in 1919 (the Americans did not start until 1926) and in its early years was the responsibility of various departments including the Department of Defence, Department of Civil Aviation and Federal Departments such as Transport and Aviation. (Keith, 1997)

The Department of Civil Aviation (DCA) was created in 1938 and controlled civil aviation in Australia until 1973, when it merged with the Department of Transport. In 1982, responsibility for civil aviation was removed from the Department of Transport, placed under a new Ministry and re-named the Department of Aviation.

Australia was a founding member of the International Civil Aviation Organisation (ICAO) and has been represented on its council since its inception at the 1944 Chicago Convention. It is also a foundation member of the South Pacific Air Transport Council. (Buddee, 1978). Australia has generally based its national requirements on the US Federal Aviation Regulations (FARs), British Civil Aviation Requirements (BCARs) and now European JARs and the differences are quite minimal. (Dunn, 1988) Reasons for not adopting regulations directly from the much larger US FAA include the concept of Grandfather rights whereby '...aircraft and their derivatives can remain in production almost indefinitely and still only be required to meet the original design standards.' (Dunn, 1988)

The 1977 Domestic Air Transport Policy Review (Department of Transport, 1979) noted that '...there had been little debate on the role of the Government in safety regulation'. Industry Commentators tended to focus primarily upon the economic considerations of the regulatory structure. Advocates for the deregulation of the domestic air transport industry suggested that 'the Australian regulatory system can be criticised for its lack of competition, minimal innovation, poor consumer choice, and high costs and fares' (Albon and Kirby, 1983; Kirby, 1979; Forsyth and Hocking, 1980; Findlay, Kirby, Gallagher, Forsyth, Starkie, Starrs and Gannon, 1984). However, minimal coverage is given to the safety implications of the existing system and any proposed shift towards US style deregulation. Findlay *et al.* (1984) acknowledge that '...statistics regarding air safety must be interpreted with care...' yet base their argument against '...exaggerated claims of some commentators that deregulation will necessarily result in a lowering of air safety standards' on the following logic.

In 1980 the total accident rate for certificated carriers was 0.221 accidents per 100,000 hours, the best result on record. In 1982 the rate was only 5% higher and the second best result of the last decade. in terms of fatalities per 100 million passenger miles flown, the average for the period 1977 to 1982 was less than half the average for the period 1971 to 1976. Similarly, in 1982 the commuter industry recorded the lowest total and fatal accident rates in the eight years for which statistics have been available.

The summary carefully ignores the general improvement in air safety that may be expected over time. This includes the effect of the introduction of human factors training from the early 1980's and the introduction of Ground Proximity Warning Systems (GPWS) in the late 1970's. As controlled flight into terrain (CFIT) remains the largest cause of fatalities in aviation, the impact of prevention strategies such as GPWS on the accident record cannot be ignored. Whilst the figures do not support any doomsday scenario regarding degradation of safety within the deregulated system, they do not paint a convincing picture of improvement either. Lowering of air safety standards does not necessarily equate to hard statistics in the form of accidents. A deeper examination may show an increase in regulatory violations and a general trend towards reduced margins of safety.

The 1977 Domestic Air Transport Policy Review also notes that within Australia, '...the safety and expedition of aircraft operations is dependent on many factors including the nature, reliability and integrity of the aviation infrastructure.' The authors add that '...because the infrastructure forms an essential part of the operation of the air transport industry, the Government, through its development and provision of most of the infrastructure, can exercise a significant degree of regulation on the industry.' Such a statement remained almost unique in emphasising the regulator's safety role when discussing economics. The experiences of the 1990's to be discussed below indicate some of the shortsightedness that seemed to go hand in hand with the focus towards economic regulation.

The Civil Aviation Authority was created in 1988 by moving airport responsibility to the newly created Federal Airports Corporation (FAC). In 1990, the CAA became a Government Business Enterprise which was tasked with covering its own operating costs. This was done through fees collected for airways and terminal use and a fuel levy for private pilots. Although the capital investment afforded through this restructuring was dramatic (increasing threefold during the first year), it was not without severe organisational shrinkage. During the first year alone $57 million was paid out in severance settlements. (Proctor, 1993) This equated to the closure of five air traffic control towers and the replacement of face-to-face weather briefings at small airports with telephone and fax links. The 1991 Review of Resources sought to streamline the CAA further by over 50% from 7,332 in 1991 to 3,641 in 1996. In the Safety Regulation and Standards Division this meant it would lose over 40% of its staff within seven months - from 727 employees to 434. (HORSCOTCI, 1995)

The repercussions of the Review of Resources (RoR) were significant and far reaching. The HORSCOTCI (1995) review noted that it '...challenged entrenched power structures and cultures within the Authority', but the effects in terms of safety were planted as latent defects in the safety system. The change in focus towards core businesses and to reduce the cost for its *customers* was to result in the revision or removal of regulations. All this occurred at a time when the organisation was enduring significant *shrinking pains* in terms of individual's roles and a general crisis of confidence.

Whilst there is general consensus that a review of resources was necessary to reduce red tape and excessive bureaucracy, '...the overwhelming reaction to the changes from within the CAA was one of uncertainty, fear, resentment and antagonism towards senior management and between CAA staff.' (HORSCOTCI, 1995)

The Air Safety Regulation Review Task force released its second report on Air Safety Regulation Issues in May 1990 (CAA, 1990) and made grave warnings about the existing Australian situation

> There are currently limited national directions in regard to surveillance procedures and methodology and Regions are left largely to themselves as to the depth and direction this activity takes. No surveillance training is carried out and the quality, scope and techniques used depends mainly on individual talents and abilities of operational staff.

The Terrell Report (CAA Board Safety Committee, 1993) subsequently observed that '...the RoR reductions were initiated at the same time as the Division and industry were facing a period of unprecedented and rapid change which required the Division to be at a peak of effectiveness and able to lead the industry through the transition period. RoR documents and the outcome of this study indicate that cuts in the Division's resources were made without regard to this imperative.' In assessing this period of the CAA's existence, the HORSCOTCI (1995) report described feelings towards it as representing '...the *dark ages* of aviation safety regulation in Australia or the *slash and burn* period of safety regulation. This was the period when (Dick) Smith was chairman of the CAA.'

On 6th July 1995, the Civil Aviation Act was amended to create two bodies from the old CAA; namely the Civil Aviation Safety Authority (CASA) and AirServices Australia. The former body is an independent safety regulatory authority while the latter is responsible for airspace management and the provision of navigation and RFFS facilities. CASA is no longer a Government Business Enterprise and is directly funded by the Government. This change, in response to the Monarch and Seaview accidents '...sent a clear signal that the Government wanted the appropriate focus given to air safety.'(Keith, 1997)

Unable to Regulate?

Two accidents in the 1990s demonstrate how the emphasis of the now defunct CAA has been found to be misplaced. Whilst both accidents had a number of contributory factors behind them, they also demonstrated similar problems with the ability of the CAA to carry out its function as regulator.

Piper Chieftain at Young, June 1993

On 11th June a Piper Navajo Chieftain struck terrain and crashed on approach to the aerodrome at Young, NSW with the loss of seven souls. The aircraft was operated by Monarch Airlines as an RPT service. The subsequent investigation by the Bureau of Air Safety Investigation highlighted numerous deficiencies within the airline and the regulator and became the catalyst for the launch of the House of Representatives Standing Committee on Transport, Communications and Infrastructure's Inquiry into Aviation Safety in the Commuter and General Aviation Sectors (HORSCOTCI, 1995).

In line with BASI's policy not to specify a primary cause, they published 34 findings and 8 significant factors which described the events behind the accident. Although the accident appeared to be a classic CFIT accident during an unstabilised approach, BASI highlighted a number of organisational deficiencies within both the airline and in the regulatory activities of the CAA.

The findings specific to the CAA were as follows (BASI, 1994);

29.   Latent organisational failures identified within the CAA included;

- a difference between the corporate mission statement of the Authority, which placed a clear primacy on safe air travel, and that of the Safety Regulation and Standards Division which appeared to emphasise the viability of the industry as its major concern
- poor planning of flight operations surveillance and poor division of responsibility
- inadequate resources, which restricted the ability of the CAA to conduct regulatory activities concerning the safety of flight operations
- ineffective communications between the local CAA District Office and Monarch
- poor control of the management of Monarch surveillance
- poor operating procedures, particularly the practice of issuing AOCs for an indefinite period.

Monarch represented an airline which was operating at the threshold of the law. On the evening in question, the crew were not checked out to fly the route and the aircraft had several unserviceabilities. However, this was not unusual to the normal operations of that airline and yet it was licensed to operate by the CAA. Gaps in the regulator's surveillance and enforcement capabilities afforded the opportunity for an airline, with a poor attitude to safety, to take advantage of the system. Whilst this says a great deal about the regulatory regime at the time, it also points to the effect of organisational culture within Monarch Airlines.

Rockwell Commander en route to Lord Howe Island, October 1994

On 2 October, Seaview Airlines Rockwell Commander 690B crashed between Williamtown and Lord Howe Island with the loss of 9 souls. Although the flight was planned as an RPT flight, it was not licensed to operate the service by the New South Wales Air Transport Council. Although '...factors that directly related to the loss of the aircraft could not be determined' (BASI, 1996), mainly due to the fact that there were no survivors, no data recorders and only a small amount of wreckage recovered, BASI did publish 37 findings. As was the case with Monarch, several of these referred to the responsibilities of the Regulator, CAA.

The Seaview accident was the last straw in terms of political and public dissatisfaction towards aviation safety. The effect of this accident was exacerbated by media attention and the fact that two of the deceased were a newly-wed couple flying out to their honeymoon on Lord Howe. In a country with as small a population and as high a value of life as Australia such events have a major effect.

Significant findings that demonstrated the problems that the CAA were suffering included:

- Airworthiness directives issued by the Federal Aviation Administration and Civil Aviation Authority did not correctly specify the engine modifications to be completed before allowing flight in icing conditions.
- The Civil Aviation Authority had not followed its published procedures in monitoring service bulletins.
- Seaview Air was not licensed by the New South Wales Air Transport Council to operate regular public transport services from Williamtown to Lord Howe Island.
- The required airworthiness-related inspections of facilities, staff and equipment were not completed prior to the air operators certificate upgrade.
- Ramp checks were carried out in response to events or breaches of regulations by the company, rather than as a check on the safety health of the company.

- The surveillance that was conducted did not ensure compliance with a number of applicable regulations.
- Surveillance checks had not been conducted since the operator was granted the low capacity regular public transport air operators certificate. However, checks were planned for 10 October 1994.
- There was inadequate follow-up of deficiencies which were identified in Seaview Air's operations prior to the issue of a low capacity regular public transport air operators certificate.
- Deficiencies were treated in isolation from those which had previously been identified.
- Civil Aviation Authority district offices displayed significant differences in their respective operating practices.
- The certification of low capacity regular public transport operators was almost entirely based on the approval/acceptance of various manuals. In this case the operator was not required to demonstrate to the Civil Aviation Authority that the Organisation and its employees would/could operate according to the standards laid down in the manuals.
- The Civil Aviation Authority had no internal procedure to review the issue of air operators certificates.

Not only was service not correctly licensed , but neither was the single pilot. The aircraft was in excess of its maximum take-off weight and had been incorrectly modified to allow continuous flight through icing conditions. (as forecast for the route on that day) As with all accidents, there were multiple causal factors, but so many of them were preventable through effective regulation, surveillance and enforcement.

Following these incidents, Phelan (1994) recorded Industry leaders' perceptions of the problems within the CAA to be;

- An unwieldy and large unenforcable rule structure which weakens the regulation, surveillance and enforcement process
- The conflict of interest between revenue-raising and standards-setting and policing functions within one authority
- Administrative and political conflicts of interest because various agencies will remain responsible to the same minister
- A need, in any new authority, for specialists with operational, commercial and technical expertise, to replace career public servants in management, and political appointees at board level.

Although considerable political drama surrounded the Australian regulator and the perceived direct effect on aviation safety, Chalk (1987) challenges the emphasis on regulatory powers in the first instance as an explanation for a high level of safety in aviation. He argues that the view that '...the regulatory standards that the state imposes on airlines and aircraft manufacturers ensures their adherence to safe design and operating procedures' do not account for the level of safety. Instead, Chalk believes in '...various market mechanisms' which, in the case of aircraft manufacture, includes product reputation. However, the structure and mandate of any regulatory body is also part of this market mechanism because ultimately it is responsible to the state it represents. In the case of the Australian CAA / CASA, Board members are political appointees and and senior management positions are decided by the Board and the Minister for Transport.

The CAA Crisis of Confidence

The House of Representatives' Inquiry into Aviation Safety in the Commuter and General Aviation Sectors (HORSCOTCI, 1995) was the consequence of a severe crisis of confidence with, and within, the civil aviation authority. The two major accidents documented above had raised public awareness of aviation safety to a very high level. In his response to the publication of BASI's report into the Monarch disaster (BASI, 1994), Minister for Transport, the Hon. Laurie Brereton '...announced a broad strategy to improve air safety regulation in Australia'. (HORSCOTCI, 1995) This consisted of two major features; namely the creation of CASA and the House of Representatives Inquiry.

However, although the inquiry made a wide range of recommendations which began to be implemented, such as through the 1997 Regulatory Review Process and the collection of safety indicators, the crisis of confidence in the ability of CAA / CASA will take some time to resolve. This process was certainly not helped by the subsequent appointment of Dick Smith as Deputy Chairman to the CASA Board and the resignation of both CASA Director, Leroy Keith and two Board members in September 1997. This followed a tactical move by Smith in the absence of the recently resigned Minister for Transport, the Hon John Sharp.

In 1996, between Smith's two appointments, Australian flight crews and air traffic controllers were asked a number of questions which related to their current feelings on aviation safety. The first asked whether they thought Australian civil aviation in general had become more safe, less safe or remained about the same.

Figure 9.1   Over the last 5 years, do you think Australian civil aviation in general has become...?

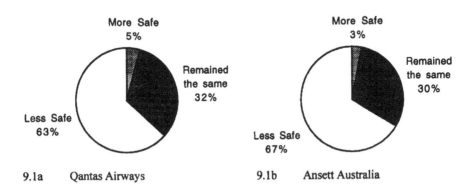

9.1a   Qantas Airways   9.1b   Ansett Australia

Qantas Airways Nearly two thirds of Qantas crew members believed that aviation safety in general had got worse over the last five years. Perhaps more importantly, only 5% believed that it had become more safe. The sobering aspect of this statistic is that with traffic increasing at current rates and the general accident rate staying static since the early 1970's aviation needs to become safety to keep the absolute number of accidents the same.

Ansett Australia A slightly larger proportion of Ansett Australia crew members (67% as opposed to 63%) believed that safety standards had slipped in the last five years in general. Only 3% perceived any improvement with a third believing safety levels had remained about the same. This represented a period of time that was long enough to be a realistically stable and yet short enough so as to exclude the 1989 domestic pilots dispute.

Whilst it is accepted that the strong negative reaction will be related to the mass of attention focussed on aviation safety at the time, it is an opinion expressed from within an educated group which has as much a responsibility for safety as the much maligned CAA. Even at this time, certain airline officials admitted that they had long since been by-passing the Australian CAA for advice in favour of the British CAA and European JAA.

Response to Occurrences

Accidents or incidents within Australia are investigated by the Bureau of Air Safety Investigation (BASI), now a part of the Australian Transport Safety Bureau (ATSB), an agency of the Department of Transport and Regional Development.

Until 1999, BASI operated as an aviation-only investigation body and as such, its function is in line with the recommended practices of ICAO's *Annex 13.* However, although the aims of occurrence investigation, as specified within the Annex are essentially reactive, BASI's role extended beyond that of similar accident investigative authorities overseas. BASI is actively involved in proactive safety initiatives which are not necessarily related to incidents or accidents which have occurred in Australia.

The Bureau is funded directly by the Federal Government and its Director reports directly to the Minister for Transport and Regional Development. Its function is kept separate from that of the civil aviation authority / civil aviation safety authority and BASI is not tasked with cost recovery. It is a relatively small agency employing some 80 staff who work in Central Office (Canberra) and at small field offices located in Brisbane, Sydney and Perth.

There are certain key traits of BASI which makes it unusual, if not unique on a world scale. These relate to the profile of the BASI team and the way investigations are conducted. Firstly, there are a number of psychologists employed by the Bureau. This included the man who was its Director for ten years up to 1999, Dr. Rob Lee. This has allowed them to look deeply into the human factors issues that are known to lie behind the majority, if not all, of aircraft accidents. Their attempts to move investigation from the traditional 'tin-kicking' technical investigations to a more systemic approach, not least through the help of human factors specialists such as James Reason, have brought critical acclaim from throughout the world.

However, such an approach is not without its detractors. Former President of the International Society of Air Safety Investigators, Richard B. Stone, suggested that following experience within the American NTSB '...all of us *(ISASI)* are dissatisfied with past results. Like many endeavours, human factors in accident investigation were oversold.'(Stone, 1997)

It is still the case that the NTSB is required to produce a single primary cause in each of its accident investigations. In Australia, the last use of a *cause* statement was in 1983 with *causal factors* lasting until about

1987, after which they just used *factors* (Mayes, 1997). Disagreement between the Australians and Americans continues with the Canadians positioning themselves more in line with the likeminded Australians than their geographic neighbours. The fact remains that the stable accident rate and seemingly static, if not growing, proportion of *human error* type accidents point towards the need to concentrate on human factors oriented investigations. Changing the emphasis from *a cause* to highlighting all of the *contributory factors* has significant merit in accident prevention and is far more than an exercise in semantics. The fact that it has been achieved first in Australia is perhaps indicative of the way of thinking that exists within that part of the aviation industry.

In terms of technical investigations, the AAIB and NTSB may reasonably be expected to hold more expertise and experience than BASI which has not had to investigate large scale aircraft accidents. The AAIB has world-class expertise in fire, severe impact and bomb investigations following accidents in UK. The NTSB has conducted a series of large scale investigations which have included difficult to access sites such as the Eastern L-1011 and Valujet DC9's in the Everglades, the Birgenair B757 and TWA B747 in the Atlantic. Australia has traditionally managed to overcome its lack of experience by sending investigators to work with overseas teams on technical investigations. For example, investigators were sent to the Dryden F28 accident, TWA800 B747 accident and the Garuda A300 crash. BASI's expertise in technical areas such as CVR analysis is routinely used around the world which is not only a reflection on their technical ability, but also adds to their operational experience.

It is worth noting the level of response from the Bureau to key incidents, which may seem out of proportion compared to the reaction of their counterparts overseas. This may, arguably, be the result of a lack of major accidents and therefore a surplus of resources. However, as the Director must justify the budget on an annual basis to the Minister for Transport it is unlikely that large-scale incident investigations are simply about finding investigators things to do.

A good example is the investigation into the accident involving Ansett Australia Boeing 747 VH-INH at Sydney Airport on the 19th October 1994 (BASI, 1996b). The investigation lasted nearly two years and culminated in an 86 page report which examined issues ranging from the aircraft and flight deck to organisational issues within Ansett. The depth of the investigation was unprecedented except by the fatal accident investigations. The consequences were similarly far reaching, especially within Ansett where the safety Department swelled from one to ten staff members and was accompanied by an intensified commitment to human factors training which will see all employees go through some form of training by 2001.

Luck - The Other Explanation

In spite of overwhelming evidence to the contrary, a recurrent theme in the anecdotal explanations of Australia's good record for airline safety is that of luck. Numerous individuals, who are asked to comment on the reasons behind Australia's zero fatality record for jet aircraft travel, used the word 'luck'. Even Dick Smith, former Board Chairman of the Australian CAA (1993) gave *luck* as one of the four reasons he believed to be responsible for the success.

Luck is defined by the Oxford English Dictionary as 'supposed tendency of chance to bring a succession of favourable events or good fortune'. Chance is not a scientific concept and cannot be defined. Its closest scientific equivalent is probability which can be estimated using past samples of similar circumstances and it is this probability that is used in aviation safety to predict accident and incident rates and types. However, usage of the term *luck* is often as a catch-all to explain the unexplained or even shrug responsibility by claiming an action to have been influenced by forces well beyond human control. Asked to suggest factors that they believe had been responsible for the good safety record in Australia, the response are summarised in Figures 9.2 and 9.3.

For pilots, luck was ranked in 5th place, having been cited by approximately 29% of crew members. However, the results from air traffic controllers were initially a little more concerning.

The perception held by military and civilian air traffic controllers is that luck has been the number one factor in preventing a fatal air accident within Australia involving an RPT class aircraft. This may represent a worrying mindset or reveal an interesting cultural trait, which is worth further examination.

Figure 9.2   **What factors have contributed to Australia's safety record? (large transport pilots)**

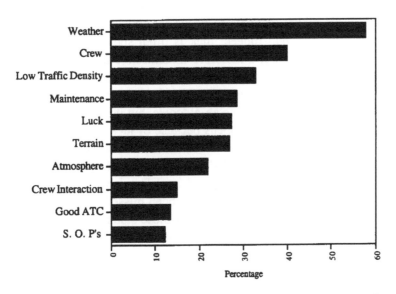

Figure 9.3   **What factors have contributed to Australia's safety record? (air traffic controllers)**

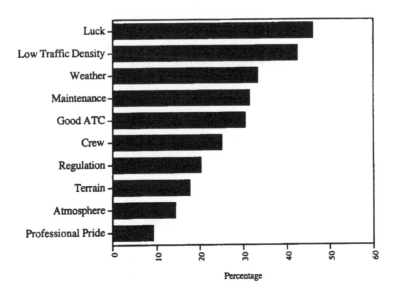

Certainly a belief in luck is not something encouraged by anyone other than the likes of Casino owners or Bookmakers. Loughborough University's Transport Professor, Norman Ashford observed in 1993 that, '...a PhD thesis which attempts to prove the existence of luck would be scratching stony ground.' Indeed, luck is often spoken of, but is entirely unprovable as a concept and therefore inadmissible as a scientific notion. The euphemistic use of the term 'fortuitous' to describe the chance outcome of a particular set of circumstances may just be an excursion into semantics. The term may be used to evaluate an *outcome*, but it is not a valid *explanation* of the process. Nothing happens because of luck and as it represents a value judgment, one man's good luck can easily be another man's bad luck:

Suppose a concrete slab dislodged from a motorway bridge and fell onto a passing motor car, injuring the driver. The driver of the car in front may consider himself to be lucky to have been missed, whereas the casualty may feel that he was unlucky. Both of these statements, as value judgments stand as valid, but neither explain why the slab fell where it did. The circumstances behind the collision may be the result of a variety of factors including erosion forces on the bridge slab, wind strength and direction, vibration from passing vehicles, the speed and direction of the vehicles passing beneath the bridge and its maintenance program.

Admittedly, many of these factors are difficult to quantify and the drivers would not be expected to readily evaluate the relative contribution of each of them as they drive their cars. However, the example does illustrate that none of the forces fall under the description of luck. In other words, luck does not make things happen, it just describes an individual's satisfaction of the outcome.

The Origin of Perceived Luck

The widespread use of the term luck raises two interesting questions for research, regardless of whether it is proven to exist or not. Firstly, If so many people cite luck and it does not exist, what do they really mean? Secondly, if we accept luck belief, what are the effects?

If so many people cite luck and it does not exist, what do they really mean?
Explaining the 'luck' factor is not easy as being a generic, catch-all term, it may mean different things to different people. One possibility is that luck is

a convenient way of explaining the unknown - which in turn may be the result of a number of factors including naïvety, ignorance, misapprehension or mistake. In the aviation context this may be because many people have a limited view of aviation safety and the factors that have an influence upon it. Simplistic attitudes which are formed on the basis of limited experience or exposure to selective information have created a number of myths about safety. For example, there is a sizeable populous within the industry which looks at accidents solely in terms of primary causes and therefore believe that pilot error is currently the most important variable in aviation safety. The progression seems to be that the failure of an airline safety system through pilot error is unlucky and not the result of the myriad of processes which systemic investigation reveals.

Luck may also be used as a way to describe the effect created by greater margins of safety. In turn, these margins may be shown to be either natural or intended - for example, clear visual conditions may afford a margin of safety which could prevent a mid-air collision following a breakdown of separation (natural) or undeclared runway pavement in excess of ICAO minima may prevent an overrun (intended). Aviation safety has always been based on fail-safety and duplication, which creates a margin of safety within which aircraft can still operate without incident. Of course, once a margin of safety is entered there is less room for error without an incident occurring so the bigger the margin of safety, the lower the probability of exceeding it.

While a number of 'near-misses' are heralded as examples of *Australian luck*, there are plenty of similar stories from around the world to balance them out. When an Ansett A320 had to make a go-around at Sydney Airport in 1991 to avoid colliding with a DC10 at the intersection of the two main runways, there were critics who suggested the difference between a near miss and an accident was luck. The Ansett flight crew would dispute that, as would the regulator who had deemed that SIMOPS (simultaneous operation of runways) should only operate when aircraft were in clear visual range of each other.

Both aircraft were landing; the DC10 on Runway 34 and the A320 on the crossing Runway 25 which was allowable under SIMOPS in visual conditions. The DC10 was required to have the other aircraft in sight and stop short of the runway intersection whereas the A320 was just required to keep the DC10 in sight throughout the landing. As the two aircraft neared the intersection, the Captain of the A320 observed that the DC10 did not appear to be slowing in time to stop and elected to go-around only a matter of a few feet above the runway. There was no collision but subsequent investigation suggested that the two aircraft may have impacted if the go-around had not been expedited.

Despite there being a number of complex issues surrounding the incident, the factors which prevented a collision could not be put down to luck. Both aircraft were aware of each other's presence and meteorological conditions were required to be clear before the mode of operation could commence. The Captain of the A320 was observing the DC10 at all times and was able to apply the go-around power as necessary. The accident was avoided, in spite of multiple failures because of the safety net provided by the regulator and airline operating systems and not because of luck.

Systematic investigation of any aircraft incident will conclusively prove that luck was not a factor, either in a positive or negative sense. The contribution of chance to the way that multiple factors came together on a particular day is not disputed, but where 'chance' is the correct term to use, there is no reason to believe its distribution to be anything other than random. For example, say an engine has a design failure rate of 1 in 50,000 hours then the likelihood of it occurring when a particular aircraft is flying on a weekday or a weekend may be reasonably classified as chance. In this example, there are no possible factors which would specifically dictate where that chance occurred. However, if such a failure occurred more frequently in a particular geographic area then deeper investigation may reveal it to be the result of a number of factors including;

- distribution of that aircraft or engine type
- level of flying completed in that area
- performance stressors e.g. temperature, precipitation
- maintenance integrity
- pilot airmanship
- use of derated take-off power.

Again, none of these factors are explained by luck or chance. This enhances the argument that regional variations in performance are not down to unknown factors which can only easily be described as luck.

Several aviation accidents and incidents have been labelled as being the result of good or bad luck. One famous example is that of *United Airlines Flight 232*, which experienced an uncontained failure of the number 2 (tail) engine in 1989 with a loss of all hydraulic power. Technically it was an unsurvivable accident, but the flight crew managed to successfully crash land the aircraft at Sioux City, Iowa with 185 survivors. The perform- ance of the flight crew was exemplary and assisted by a 'dead-heading' Check Pilot who administered the crucial variable thrust (the only remain- ing control on the aircraft). Was the fact that the Captain of the flight had spent some of his own time practicing such a scenario in the flight simula- tor beforehand a stroke of luck? Was the presence of a Check Pilot another stroke of luck? In both cases, it seems that there was some fortuity, but there are also a number of issues that were definitely not luck including the atti- tude of the crew members towards extra-curricular training, access to the simulator and the use of scheduled services to dead-head crew. To consider the outcome of this incident to be the result of luck detracts from the efforts of the crew and airline.

If we accept luck belief, what are the effects? There is no reason why Australian carriers should have been blessed with more luck than those from any other nation. However, this has not prevented the development of the widely held belief of the importance of luck in creating the safety record.

As mentioned earlier, there are a number of people who put Australia's good safety primarily down to luck. There is no doubt that there have been a number of incidents where an accident was very narrowly avoided by circumstances which could easily be described as luck, but it is not responsible for the safety record of a nation. There are those who say that explaining factors which others may have put down to luck is an excur- sion into semantics, but this ignores another crucial issue: If the next gener- ation of aviators are brought up to believe that luck has played a major role in the safety record, then there is a danger that they will believe this force to be beyond their control. In other words, the danger is either in taking the complacent approach of believing Australia to be the 'lucky country', or the fatalistic approach of assuming luck to be out of human control and there- fore will inevitably run out at some point.

Trompenaars (1993) looked at indicators which could assist in the quest to determine whether luck believers existed because of cultural traits. Asked a question to test whether individuals believed themselves to be Captains of their own fate, the following table charts the percentage of respondents who believe that what happens to them is their own doing;

Table 9.2   **Percentage of respondents who believe that what happens to them is their own doing**

East Germany	35	Italy	65	Netherlands	77
China	35	Hong Kong	69	Ireland	78
Egypt	48	Romania	70	Norway	80
Japan	56	Sweden	70	France	81
Turkey	57	Finland	70	Spain	81
Czechoslovakia	57	Thailand	71	**Australia**	**81**
Singapore	58	Brazil	71	Argentina	81
UAE	58	Austria	71	West Germany	82
Nigeria	59	India	72	Canada	83
Poland	62	Indonesia	73	Pakistan	84
Greece	63	UK	75	Switzerland	84
Portugal	64	Belgium	76	USA	89
Ethiopia	64	Denmark	76		

Source: Redrawn from Trompenaars, 1993.

A very high proportion of Australians (81%) believe they have significant control over their own destiny which in in contrast to some of the more collectivist cultures. Whilst this phenomena is partly related to the position individuals believe they hold in society, it is also related to the number of controllable variables they perceive. Where these variables are not controllable, especially under collectivism, success may be perceived to be achieved through luck or other factors beyond the individual's control.

Notwithstanding this, the overwhelming evidence in Australia is that aviation safety and its safety record for airline operations is not a function of luck. However, chapter 10 discusses why so many people seem to be strong believers in luck as the determining factor.

10 Exploring the Operational Environment

Aircraft Flying Characteristics and Performance Limitations

Differences in aircraft design will determine the performance level and ability to cope with certain operating excesses. The operation of aircraft close to their intended performance envelope reduces the margin of safety. Conversely, an operator that specifies extra equipment or capabilities for a particular aircraft model may also reap the benefits of a wider margin.

Structural failure of aircraft is a rare event and one that is usually associated with poor maintenance levels rather than airframe design. A number of accidents occurred during the first years of jet aircraft when aircraft were flown beyond their capabilities. However, in the majority of such cases, the primary cause was one of poor training or crew overconfidence, rather than poor design.

Certain aircraft types are prone to airframe icing which has serious flight safety implications. The low wing, rear engined DC9/MD-80 family and Fokker F28/70/100 aircraft are all prone to clear icing on their wings. When shaken loose, particularly by take off, ice may be ingested by the engines with a danger of loss of thrust at the most critical point. Accidents occurred in 1991 (SAS, MD-81) and 1989 (Air Ontario F28) with this as a contributory factor. The loss of an ATR 72 in 1993 led the NTSB to conclude that the aircraft had undergone an uncommanded assymetric aileron excursion caused by a build up of ice on the wing.

As documented earlier, in 1992, an Ansett Australia BAe 146-200 aircraft lost power in all four engines while flying around a large storm cell near Meekatharra, WA. Flying into an area of unusually warm air at high altitude, the engines were unable to supply sufficient fuel and power was lost through *roll-back*. (BASI, 1994c) Although the manufacturer was aware of the possibility of *roll-back* and had issued an operational notice, there were a number of omissions such as the possibility of more than one engine being affected. Following the incident, new operational ceilings were set for the aircraft type to prevent recurrence.

Unforeseen Type Deficiencies

New aircraft types are only certificated for use after extensive flight trials. However, in certain cases, aircraft types have been certificated with latent design defects which later present themselves through incidents. Whether such oversights are a function of imperfect regulation or simply a limit to the level of knowledge is of little consequence to the operator who suffers the fault. By definition, such accidents are beyond the control of the operator, but nevertheless can significantly affect an airline's safety record.

The accident record reveals that such incidents are few and far between. The fact that the majority of the world's aircraft are built either in USA or Europe where regulatory standards are very high make unforeseen deficiencies a rare event. However, the series of accidents involving DeHavilland Comet aircraft illustrated how advances in technology can still fall foul of the unknown. The Comet 1 was built with square windows which allowed fatigue cracks to develop under the stresses of pressurised operation and resulted in a number of catastrophic structural failures. The loss of three BOAC Comets led to the withdrawal of the aircraft's certificate of airworthiness and also significantly affected the airline's safety record.

There is also debate as to the unpredictability of the DC10's design fault regarding the rear cargo door. In 1972, an American Airlines DC10 flying over Windsor, Ontario experienced a rear cargo door blow-out, sustaining damage to its control systems which were all directed through the cabin floor to the tail of the aircraft. (Stewart, 1987) The aircraft landed safely and modifications were recommended to the latch mechanism of the door. However, it seems that the manufacturer's modifications to new aircraft were incomplete and in 1974 a Turkish DC10 was lost at Ermenonville, near Paris after a similar cargo door blow-out.

Barnett *et al.* (1979) suggest that this loss was not to be classified as a design flaw as '..apparently it was at least somewhat related to questionable maintenance procedures by the airline involved'. The modification carried out by Douglas to the latch was defective and made the likelihood of a failure higher. It meant that a little amount of brute force applied by a ground handler could cause the pin to bend, appearing to be in position when the opposite was true. As ground handlers are generally recruited for brawn over brains, some of the responsibility must lie with Douglas. In other words, had another carrier been operating that particular aircraft, there is a significant chance that the same accident would have happened and so it is a poor indicator of the airline's attitude towards safety.

Existence of Hazard in Australia

Significant design problems are reviewed below for the large transport type aircraft which may have or do operate within Australia.

DeHavilland Comet   The first version of this aircraft was not ordered by Qantas as they were already happily equipped with Lockheed Constellation aircraft for long distance routes. However, with the takeover of British Commonwealth Pacific Airlines, Qantas inherited an order for Comet 2s '...subject to the usual technical considerations' (Gunn, 1988), but as the Comet disasters saw the Airworthiness Certificates revoked for Comet 1 and 2 aircraft, the order was never completed.

The Comet 4 was considered by Qantas in competition with the B707, L1649 Super Constellation and Bristol Britannia 310. One of the concerns for Qantas was the poor record of structural integrity experienced by DeHavilland aircraft. Dick Shaw, technical advisor to Qantas (cited in Gunn, 1988) assessed the Comet 4 and although he commented that there was no trouble or expense spared in ensuring the aircraft was free from structural problems,

> The integrity of an aircraft design depends on the philosophies and objectives of the Chief Designer and his section leaders. It also depends on the integrity, ability and outlook of all the design staff, down to the most junior design draughtsmen. It is quite impossible for an outsider, no matter how technical, to assess a large firm for the integrity of its design detail on a new and unproven project. In this respect the common and prudent practice is to judge each manufacturer on its past record.

Qantas chose to order the B707 and no Comets ever flew in Australian service. History shows that the design related accidents were limited to the early series 1 aircraft. Although the second production batch (Comet 4) recorded a significantly higher accident rate than more recent designs, in terms of total losses, it had a better record than the DC8, SE210 and B707/720.

Douglas DC10   None of this aircraft type have ever been purchased by Australian carriers. In 1971 Qantas considered the DC10 alongside a further order of B747 aircraft, but chose the latter option on grounds of price, operating costs and type commonality. Had the airline bought the DC10 then it is not inconceivable that a Windsor-style incident could have occurred. However, it is debatable as to whether a Ermenonville type accident may have befallen Qantas. Following the Windsor incident, certain airlines (e.g. Laker) took the initiative to fit additional fail-safe locking devices to their aircraft. It seems likely that Qantas would have fallen into this category of airline.

Boeing 767   Both Qantas and Ansett operate this type of aircraft. If the findings of the accident investigation are to be believed, it would be possible that the in-flight deployment of reverse-thrust equipment suffered by Lauda Airways in 1991 could have occurred to an Australian aircraft with similarly catastrophic results. The discriminating factor which caused the accident to happen to a Lauda aircraft is not known. All B767 aircraft were modified at the instruction of the manufacturer following this accident.

Boeing 737   At the Australian House of Representatives Inquiry into air safety, a CAA airworthiness inspector revealed that an Australian B737 was allowed to fly on domestic routes for four months after the similar fault which may have caused the loss of the Lauda Air B767 was discovered (Herald Sun, 1995). The warning light had been disconnected and although the CAA had been led to believe that the problem had been rectified after a few days, a subsequent inspection revealed this not be the case. The inspector blamed '...unqualified or incompetent people making judgements about the plane's safety' for the occurrence. As this incident followed the Lauda Air crash, then responsibility must fall solely on the operator for ignoring a design defect that was now known of and could be fixed.

Debate following the loss of two B737s in separate unsolved accidents is cause for some concern. In March 1991, a United Airlines B737 crashed on approach to Colorado Springs with a loss of 25 souls. This was followed by a similar accident in September 1994 involving a USAir B737 on final approach to Pittsburgh with a loss of 132 souls. Both accidents remain unsolved although the main theory centres around an uncommanded rudder excursion which pitched the aircraft's nose to the ground in an unrecoverable dive. Rudder deflections have occurred to British Airways

and Qantas B737 aircraft, albeit without such catastrophic consequences.
BAe 146 The engine roll-back incident at Meekatharra (BASI, 1994c) represented an all engine failure scenario that had not been anticipated by the manufacturer or operator. The aircraft involved lost significant altitude (21,000 ft) before engines recovered. This represented a very close call for the industry and operating restrictions were placed on all BAe 146 on the Australian register.

Whilst the above aircraft types have been highlighted as suffering from major problems, it should be noted that all aircraft designs will accumulate defects which will need attention through airworthiness directives or manufacturers' advisories. These include modifications such as airframe strengthening, avionics upgrades or engine alterations. This is a standard process within the industry which reflects the considerable investment represented by aircraft. Airworthiness directives must be complied with, but advisories are at the discretion of individual operators and this provides an indication of corporate attitude. All of the aircraft types operated by Australian carriers have had airworthiness directives issued. This does not necessarily suggest that the aircraft were unsafe, rather that modification improved their operational efficiency.

Another consideration is the fact that *unforeseen type deficiencies* can strike at any time and there is no reason to believe that this will not continue to be the case in the future. Recent accidents which are still under investigation such as TWA 800 off Long Island in 1996 may point to serious problems with the centre wing fuel tanks of B747 Classics; an aircraft type operated by both Ansett and Qantas. Further, there is always the risk of problems associated with ageing airframes such as the catastrophic airframe failure suffered by an Aloha Air B737 in 1988. Ansett, for example currently operate some of the oldest B767s in the world.

Aircraft Maintenance

The maintenance of aircraft is fundamental to the safe operation of the aviation system, yet its impact on the accident record is often underestimated. Boeing (1996) report the proportion of *Maintenance as primary cause* accidents to be comparatively low, especially when compared to *Flightcrew as primary cause*.

Boeing records an average of 3.4% Maintenance Primary Cause Accidents for the *jet age* period 1959 and 1995 which rises to 4.9% in the ten year period 1986 to 1995. Ashford (1994) also records a low proportion of maintenance causal factors in his analysis of 219 accidents with only 15 maintenance factors highlighted out of the total of 839. However, it is also possible that other factors listed, such as engine failure or structural corrosion may be classified as maintenance related.

Whilst this proportion of primary cause and causal factors may seem low, other research has highlighted the fact that 'maintenance incidents contribute to a significant proportion of worldwide commercial jet accidents' (BASI, 1997). Hobbs and Williamson (1994) examined human factors in airline maintenance and highlighted the fact that in terms of fatalities, maintenance accidents accounted for the second highest number of fatalities after controlled flight into terrain.

The authors cite figures prepared by the Boeing Safety Engineering Department for the ten year period 1982 - 1991 which attribute 1,481 fatalities to *maintenance and inspection* issues. Such statistics include accidents with loud statistical noise such as the loss of a JAL B747 at Mount Osutaka in 1985 which claimed 524 lives; the highest toll in a single aircraft accident.

BASI (1994b) also note that Marx and Graeber (1994) '...estimated that 12% of major accidents involved maintenance as a contributory factor.' Reason (1994) also notes a study by Weiner of Boeing which examined 200 fully documented accidents and found maintenance and inspection errors present in 34. Whilst maintenance deficiencies rarely the primary cause of accidents, they are much more frequently found as contributory factors. Accidents may occur as a result of improper handling of the aircraft in response to a *survivable failure*, but in these circumstances, it may be argued that the failure of the last line of defence would not have been catastrophic if it had not been needed in the first place.

In spite of the significance of maintenance failures in aviation safety, '...there is a lack of empirical research on the nature of maintenance incidents and the human factors which contribute to them' (BASI, 1997). 'Yet maintenance errors continue to account for a small but conspicuous number of accidents' (Reason, 1994). One of the most important issues to come to the fore in maintenance safety is the predominance of human error. This is a fact that can often be lost when *pilot error* and *human error* become used interchangably and *maintenance error* is left as a separate category, almost as if the maintenance side of aviation is not a human centred function. For example, Stone and Babcock (1988) ask how to address the issue of pilot error, stating; 'The labels *pilot error*, *operator error* and *human error* all describe the same characteristic.' This is far from the truth as pilot error refers specifically to errors made by (or blamed on) the pilots whereas human error is the larger subset which includes errors made by pilots as well as others such as maintenance workers, designers, air traffic controllers and management. To give the impression that, in aviation, human error is exclusively the domain of pilots is, at best, old fashioned and, at worst, dangerous (Braithwaite, 1997).

Using figures produced by the UK CAA (1992), which looked at 6,672 occurrences to UK registered aircraft over 5,700 kg Maximum Take Off Weight (MTOW), Hobbs and Williamson (1994b) observed a total of 537 maintenance discrepancies. Of these, the top eight factors were as follows;

Table 10.1   Top eight maintenance discrepancies

Discrepancy	Percentage
incorrect installation of components	50 +
wrong parts fitted	6.7
electrical wiring discrepancies	6.1
loose objects left in the aircraft	4.8
inadequate lubrication	4.8
cowlings / fairings not secured	4.1
fuel / oil caps not secured	3.9
landing ground lock pins not removed	1.7

Source: Redrawn from Hobbs and Williamson, 1994.

Although these statistics relate to incidents as well as accidents, they illustrate the prevalence of human error within the maintenance environment. Indeed, having established the importance of maintenance at an industry level and the centrality of human error to the frequency of maintenance or inspection failures, it makes sense to examine the Australian system in the context two areas; firstly, whether maintenance error is prevalent but mitigated through other lines of defence such as pilot performance and secondly, if maintenance error is less prevalent, how the human factors issues have been controlled?

How Common are Maintenance Failures in Australia?

Anecdotal evidence regarding the reasons behind Australia's good record for airline safety cited the quality of maintenance as a significant factor. Both expert witnesses and those involved in the survey suggested that the quality of maintenance was good. Figures 9.2 and 9.3 (page 196) show that both Australian Pilots and Air Traffic Controllers ranked *good maintenance* as fourth place behind weather, crew factors and low traffic density; and luck, low traffic density and weather respectively.

These perceptions are very important, as pilots in particular represent the end users of the maintenance product. A significant number of expert witnesses also highlighted the quality of the maintenance culture within the Australian carriers. It is typified by the attitude towards minimum equipment list (MEL) failures. MEL's can be *carried-over* on aircraft, subject to approval in the aircraft manufacturer's and airline's master minimum equipment lists. The aircraft is still deemed to be airworthy, albeit perhaps with less safety systems available. The use of MEL failures is to allow aircraft to continue on service flights back to their maintenance base. Ansett, Australian and Qantas strictly believed that aircraft should not carry MEL deficiencies over between maintenance checks which is common practice elsewhere in the world. However, there is some concern that this situation is changing as a result of commercial pressures and may therefore become a safety issue in the future.

A strong emphasis on engineering reliability has existed within the Australian aviation industry from its early days. This was largely a function of the tyranny of distance and the characters involved in attempting to conquer it. The former concept meant that aircraft needed to be imported into and assembled in Australia and the latter demanded that aircraft were reliable enough to fly between disparate locations over inhospitable terrain.

Qantas is one of a very small number of airlines that actually assembled its own aircraft and as such developed an exceptional engineering knowledge. In 1926, under the direction of Arthur Baird, Qantas assembled six DeHavilland DH50s, something which 'Hudson Fysh believed was the company's most remarkable achievement' (Bunbury, 1993). Whilst this procedure ceased well before the jet age, it established a tradition of engineering excellence which is deep rooted in the culture of Australian aviation.

A factor which has remained important is the distance Australia finds itself from airframe and engine manufacturers. It is a country which has never built high capacity turboprop or jet aircraft and depends on manufacturers situated in Europe and North America. This is one of the reasons why the Australian industry has a high degree of self sufficiency when it comes to aircraft maintenance. Qantas '...has the largest aero-engineering organisation in the southern hemisphere, with a completeness of stocks and engineering skills which owes its qualities to Australia's remoteness and need for independence' (Learmount, 1987).

The distance factor has had a dual effect, not just in terms of remoteness from manufacturers, but also in the need for reliability. Sparse population and poorly developed land transport systems meant that emergency landing sites were few and far between. Terrain and weather conditions which may be considered to be relatively benign by modern standards, were downright treacherous in the early days and aircraft reliability needed to be high. The high profile role played by engineers such as Arthur Baird in Qantas helped to develop an organisational culture which was not dominated by pilots as so often was the case in other carriers.

How is Human Error in Maintenance Controlled?

Human error is inherent in any socio-technical system and aviation maintenance is no exception. Assuming that Australian carriers fly similar aircraft to their overseas colleagues in similar conditions, it is reasonable to also assume that their engineers are open to similar opportunities for error. The fact that there have been no fatal accidents involving Australian RPT jets points to a high level of maintenance integrity, and, as was the case with the absence of pilot error, this in turn points to control of human error in maintenance.

Many of the problems faced by maintenance engineers have parallels with the experience of flight crews. Errors fall into two categories; errors of commission and errors of omission, shifts are long and generally during unsociable hours. However, it is an area which has often been ignored in favour of the flight crew. Whilst the most frequently quoted statistics regarding human error tend to focus on pilot error (see Beaty, 1996; Hawkins, 1987; Hurst and Hurst, 1976; Stone and Babcock, 1988), a deeper examination of maintenance primary cause accidents would reveal that the vast majority, if not all, of those incidents involve human error.

In an attempt to minimise the effect of error (and acknowledging that it can never be totally removed from the system), several strategies exist within Australia at both the conscious and subconscious levels. Major carriers have always run intensive engineering apprenticeship schemes. In Qantas, this represents an annual cost of around $7.5 million which is supplemented by a further $1 million spent on engineering skills training for permanent staff. Whilst Licensed Aircraft Maintenance Engineers (LAMEs) are licensed by the Regulator, they are also expected to pass exams within the airlines which are set at an even higher standard.

Quality Assurance Departments exist within both Qantas and Ansett which operate well in excess of the minimum requirements set by the Regulator. These formal structures are supported by other mechanisms which include the Qantas HEAR (Human Error and Accident Reduction) group and Engineering and Maintenance Incident Investigations. Qantas has also had the post of *Engineering & Safety Manager* for a number of years who is a Safety Department employee with functional responsibility in the Engineering & Maintenance Divisions.

However, in spite of formal quality assurance and training systems, it is still possible for dangerous errors to occur if the culture in which they try to operate is deficient. A poor attitude towards quality assurance systems will see them circumvented (for example, see Clapham Junction Railway Disaster, 1988). Two of the reasons that Australia's safety record is good are corporate attitude and an openness in communication, both of which have been discussed earlier in the text. In the maintenance environment, this translates into an attitude that believes things must be fixed promptly and fixed first time. It also means that mistakes can be admitted and corrected without fear of reprisal or loss of face, and that errors can be highlighted with out animosity. Young (1988) observes that '...there is no way in which Australian LAMEs will meekly do as they are told. They will do a task which is assigned to them, but will only do it if they are confident that the job can be accomplished safely, and the subsequent result is the correct one.'

Buyer Furnished Equipment

The provision of secondary safety equipment on board an aircraft to miti-gate the consequences of an accident is controlled to a degree by the regu-lators and aircraft operator. Some equipment such as seatbelts and over-wing exits come fitted as standard by the manufacturer, whilst others are termed 'buyer furnished equipment' (BFE). The latter category may include life-rafts, medical bags and defibrillators. The supply of buyer specified equipment can significantly improve survivability in incidents and acci-dents.

Due to the nature of buyer furnished equipment, the number of cases where accidents have been avoided because of their existence are poorly reported. Examples where equipment such as medical bags and defibrillators have had a positive benefit exist at company levels, but are rarely recorded elsewhere. A few anecdotal examples do exist, the most famous being an emergency operation carried out on board a British Airways Boeing 747 aircraft *en route* from Hong Kong to London in 1995 using a coathanger, catheter and water bottle.

The discussion surrounding the value of some on-board equipment such as lifejackets and rafts compared to potential new equipment such as cabin water sprays is a difficult one. Ditchings at sea involving large jet air-craft are very rare and the limited experience that is available seems to demonstrate that the majority of fatalities are the result of impact forces and would not be mitigated by the use of life-jackets. The important question is, of the survivors, how many would actually have done so regardless of the provision of life-jackets. This is an important question for future research.

The long distance intercontinental operations conducted by Qantas required aircraft to be flown to the limits of their range capabilities. Therefore, any excess weight would have a significant effect on operating costs and should be eliminated. The advent of the Boeing 707 jet brought new possibilities in endurance flying, but required tight controls on weight to be economical. Evacuation slides had become mandatory for aircraft of that size (not least because the new breed of aircraft had sill heights 16 feet above the ground.) and cross water operations required the carrying of life-rafts for use in the event of ditching.

With the Qantas B707-138 series aircraft flying some of the longest range routes in the world, any weight penalty would have hit the airline hard and seriously affected operational profitability. One of the Qantas Engineers came up with an idea that was destined to become industry standard. This was the design of a a combined *slide raft* which was as useful for evacuation as it was for ditching and saved the weight of duplication. The invention was perhaps more significant than simply being a good idea. Qantas came up with a revolutionary design and gave it to the industry for nothing. It may seem obvious that any airline would be focussing its attention on maximising the available payload, but Lewis (1995) believes that this is a good example of innovative *pioneering spirit* which exists with the Australian aviation industry.

Security Equipment

It is a legal requirement that all RPT aircraft entering Australian airspace must carry handcuffs or similar restraints for the control of unruly passenger. Although problems of this nature are a relatively recent phenomena, the mandatory carriage of such equipment has not yet reached Europe. British charter carrier, Britannia who operate to a number of worldwide destinations, only carried restraints on the aircraft operating to Australia. Interviews with training staff reveal that the restraints have only been used *once or twice* during the three seasons of operation and would be of much greater value on European operations such as to Ibiza and Tenerife.

      Ansett and Qantas flights carry restraints by law and have been used on a number of occasions. In particular they seem to be need on the long night flights from Perth to Sydney or Melbourne and on "footie specials" (In Australia, distance forces football fans to travel by aircraft to certain games and any hooligan element, albeit diminished by the expense, are shifted from trains to aircraft.) It is further anecdotally speculated by flight attendants that the higher than average proportion of male flight attendants to be found at the Australian carriers leads to a greater willingness to intervene with boisterous passengers.

Heavy Crewing

The B747-400 series aircraft brought not only added capacity to the jumbo jet market, but also a new extended range capability. Routes such as Los Angeles - Sydney (14h 20m) and Singapore - London (13h 40m) could be accomplished in a single sector with an added operational bonus that Boeing had designed the aircraft for two-man operation. Although not all B747-400 customers wished to use the full potential of the aircraft's extended range, long distance operators were faced with the problem of crew duty time over flying time. Qantas has always maintained a policy of *heavy crewing* (Green, 1993) whereby additional crew members are carried on flights which might legally be able to be crewed with less crew, albeit at the extremes of their operating capabilities. Rest bunks are fitted for use by the additional technical crew members as BFE.

Quick Access Recorders

The Quick Access Recorder (QAR) is a form of flight data recorder which allows the operator to remove a tape or disk following a flight to monitor the performance of both the aircraft and the crew. Used correctly and with the co-operation of both pilot unions, the QAR information can help;

* *Monitor the operating standards of the airline directly to identify any undesirable trends that may develop on a daily basis*
* *Make sure that training and route check procedures are achieving their objectives on the line and adjust them if the Flight Data Analysis program shows any problems or trends*
* *Provide continuous monitoring of aircraft systems & performance*
* *Monitor any engineering problems that may be affecting the operation of the fleet*
* *Monitor operation requirements and/or constraints imposed by outside agencies, e.g. air traffic control.*

Source: Redrawn from Faulkner, 1991.

Qantas decided to look at operations monitoring systems in 1987 under the direction of a steering committee appointed by the Directors of Flight Operations and of Engineering & Maintenance. A decision was made to implement a continuous monitoring system on all of the company's aircraft. (which at the time consisted of B747 Classic, B747-400 and B767) The two types of system which were available were continuous monitoring or exceedance monitoring. The former requires all of the flight to be recorded for analysis at a later date by the ground station computer. The latter method requires additional on board equipment and records only exceedances to pre-programmed operational limitations.

The benefits of such a system are numerous. For example, from an operational engineering point of view, the use of QAR data allowed Qantas and Rolls Royce to more actively monitor the use of derated or reduced-thrust operations. The closer monitoring afforded by QARs has led to up to a fifty percent increase in cycles between component change. Cost savings associated with this procedure alone are enough to cover the costs of running the program. (Faulkner, 1991)

From a safety perspective, one of the early success stories of the programme came from monitoring the take-off performance of the B747-400 fleet. Qantas became aware that aircraft speed was bleeding back to V_2 and below shortly after take off, especially when aircraft were close to their MTOW. Trend analysis conformed that this was something that was occurring with different crews at different airports and further investigation was made. It seems that in the change from the B747 Classic aircraft to the glass cockpit -400 series, crews were tending to over-pitch the aircraft and lose air speed. Aircraft were coming close to suffering tail-strikes or stalling and amendments to the conversion training were made. A sharp decrease in such exceedances confirmed and quantified the success of the program.

Perhaps the most important aspect of the scheme is the way it was set up and run. This gives some insight into the attitude of the company and its crews and ultimately makes a significant difference to the long term viability of the scheme. Indeed, without the right culture within the airline, such a system could not have been implemented as smoothly as it was. Former Qantas General Manager, Operations, Alan Terrell suggested that the system was '...one of the best things that Qantas ever did' (Terrell, 1995) and that one of the critical elements was the positive role of the pilots' union (AIPA) which was unusual in '...its commitment to maintaining standards and conditions'. In other words, the establishment of such a system may well be a strong indicator of an organisation's safety culture.

The Qantas flight data analysis office is based adjacent to the safety department and operates as a clean-room environment with strict security clearance required for access. All data is handled by a small, dedicated staff who will not entertain visits from active crews to the facility, to protect the integrity of the scheme. Should a pilot make a significant mistake or deviation which is flagged by the monitoring computer then they will be approached in a subtle and non-punitive way. This is usually in the form of an informal telephone conversation with the AIPA (Australian International Pilots Association) representative who will make inquiries into whether performance exceedances were the result of a *bad day* (e.g. family problems) or a training deficiency. The union may be able to provide support in the case of the former or the pilot may chose to address the latter deficiency during recurrent training. Support for the scheme by the AIPA and its confidential, non-punitive nature are fundamental to its success.

In contrast, a similar scheme of on-condition monitoring was introduced into a certain Asian airline. The Chief Pilot of that airline explained how the system worked at a conference and was questioned about how the company responded to recurrent exceedances by particular individuals.

Q. What do you do if one pilot keeps making the same mistake?
A. We will phone him and tell him so.
Q. Do you offer any sort of retraining?
A. No, there is no demotion, punishment or retraining.
Q. But if the pilot keeps making the same mistake, do you not offer
*   retraining to them?*
A. No, there is no demotion, punishment or retraining.
Q. But what if he keeps on making that mistake?
A. He will not do so.

In other words, this was a scheme operating in a culture where retraining was seen as being similar to punishment or demotion. Crew were told of their mistakes and it was entirely up to them to correct their mistakes. The concept of recurrent mistake making was embarrassing to the culture and, to save face, was not considered in the formal response structure.

If this is the way that the QAR monitoring system works then it is likely to have deficiencies in the area of highlighting individuals' problems. However, this is not to say that the other aspects of the programme such as the engineering uses are any less effective than the Qantas system.

Collision Avoidance Systems - TCAS

TCAS or Traffic Alert and Collision Avoidance System is a radar system which detects other aircraft which may be on a collision course, using transponders which are carried on other aircraft. The TCAS system issues alerts and will suggest remedial action such as *climb* or *descend*. They are also intelligent systems such that should two aircraft fitted with TCAS equipment be flying on a collision course, they will be issued with opposing instructions i.e. one aircraft will be advised to climb and the other will be advised to descend.

The US FAA have mandated the fitment of TCAS equipment to all large RPT aircraft entering the country and Australia has recently followed suit. All of Qantas and Ansett Australia's fleets are fitted with TCAS. Interviews with expert witnesses suggest that the TCAS equipment is mostly needed for services which overfly the Far East and Eastern Europe where ATC provision often leaves something to be desired.

Mid-air collisions are comparatively rare although there have been a number of significant accidents where aircraft have collided even in clear skies (e.g. Zagreb, Cerritos, San Diego, Charkhi Dadri). As a result of the inertia involved in the collision of two aircraft, these events are usually total fatalities and are therefore particularly feared by air crew. The survey conducted during this research found that the greatest fear of Qantas pilots (international and domestic) is that of mid-air collision. This fear seems to have been heightened by an incident which occurred during the Gulf War where a Qantas B747 came within 50 feet of a USAF C5 near Phuket. As such, neither aircraft had a TCAS unit and both aircraft came extremely close - only altimeter error prevented a mid-air collision. This story remains high in the consciousness of Qantas crews and was added to by an incident which occurred near to Broken Hill in 1995 when a Qantas Domestic B737 was involved in a serious breakdown in separation with a British Airways B747. It was immediately following this incident that the decision was taken to fit the Qantas domestic fleet with TCAS equipment.

Regulatory Requirements

It is the responsibility of the Regulator to ensure compliance with minimum standards in an attempt to ensure the safety and efficiency of air transport. In Australia, this is the responsibility of the Civil Aviation Safety Authority (CASA). The basis of national standards are the recommendations set out by the International Civil Aviation Organisation (ICAO) through the Annexes of the Chicago Convention. These recommendations are then accepted with little or no modification at member state level to become law. In Australia, standards have tended to be based upon '...US Federal Aviation Regulations (FARs) and the British Civil Aviation Requirements (BCARs) and the differences from them are quite minimal.' (Dunn, 1988). Known as

Australian Civil Aviation Requirements (ACARs) they cover a range of areas which can directly impact upon the quality of an airline's operation. These include licensing of;

- air transport
- provision of accommodation in aircraft
- registration and certification of aircraft
- safety of aircraft (including airworthiness)
- certification of operators of aircraft
- licensing of flight crew and maintenance engineers.

In terms of the operational suitability of aircraft, it is the issue of airworthiness that seems to be of paramount importance. Although some definitions refer to airworthiness as the condition of an aircraft as being "fit to fly", Dunn (1988) contends that in Australia, it represents more of a process, which is put into practice through the following:

- The establishment of appropriate design and maintenance standards for the particular operation and aircraft type
- The certification of the aircraft type to the relevant design standard; including navigation systems and other operational equipment for the designed operational role
- The maintenance of particular aircraft to the relevant maintenance standards; in organisation approved for the purpose and / or by individuals appropriately licensed
- The ensurance of continued fitness to fly of individual aircraft and aircraft types through in-service safety performance monitoring,   the outcome of which can and does feed back into revised design and maintenance standards.

There are a number of reasons behind Australia demanding its own certification, rather than accepting those granted by the FAA or UK CAA as is the case in many countries with equivalent and larger aviation industries. These include issues such as *grandfather rights,* which have been accepted in the US, but not in Australia in certificating derivatives of existing aircraft series. Far from implying that other airworthiness authorities are deficient in their certification processes, the rationale behind Australia's independent certification is one of insurance or *playing safe.*

Dunn (1988) summarises the reasons as;

- A legal requirement for the Authority to be satisfied as to the airworthiness of an aircraft before a certificate is issued.
- To give freedom to manufacturers and operators to develop design, construction and operation of aircraft in the most efficient manner, airworthiness requirements, with few exceptions, are expressed in terms of basic safety objectives, rather than black and white limits. This inevitably involves interpretations, equivalent safety considerations and such. It is an essential element of conformity certification considerations to understand the basis for these decisions.
- The most important reason is that the certification process itself provides the cornerstone for the continuing airworthiness function of the Authority. It establishes the required airworthiness function it aims to maintain.

This attitude has changed somewhat in recent years as CASA has looked at harmonising regulations with those of the US FAA and European JAA (Joint Airworthiness Authorities). This followed a resolution adopted by the 29th Assembly of ICAO encouraging global harmonisation and a general attitude change within the Australian industry throughout the relatively tempestuous 1990s.

In asking whether there are examples of accidents which had been prevented as a result of Australian certification, Dunn argues that there is no satisfactory answer; 'There is no guarantee that had deficiencies discovered by the Australian teams not been corrected, an aircraft would have been lost.' However, the author does concede that 'The consequence is a lower probability of an accident occurring' (Dunn, 1988).

Nevertheless, there is a widely held belief that the style of regulation in Australia has been an important factor behind its good airline safety record. Brogden (1968) suggests '...the strict system of safety and operational controls' to be a significant factor and Ramsden (1976) also refers to '...Australia as a "police state" in its air safety regulation.' The reasons for this approach reach beyond the way that ICAO recommendations were interpreted. The HORSCOTCI report (1995) observed that; 'Other influences on regulatory development include requests from industry, community groups, and the public, government directives, international Airworthiness Directives from manufacturers or government agencies; major defect reporting systems, and the results of surveillance activities.'

Regulation of aviation safety is not a simple process in that there is not one single level of acceptable safety for all classes of operation. To set such a level would be all but impossible, not least because of the prohibitive costs that would be involved at the lower capacity end of aviation. Smith (1992) notes that 'A large aircraft could be built with the speed, comfort and safety characteristics of a Cessna 172 which would be cheaper per seat to construct and operate. Conversely, an aircraft with only four seats could be built with the same level of safety and other features as a Boeing 747, but it would cost tens of millions of dollars and nobody could afford to travel in it.'

In line with ICAO standards, regulations within Australia prescribe air safety standards according to a hierarchy of classes of operation. (HORSCOTCI, 1995) The highest standard is for high capacity RPT aircraft; scheduled services for fare paying passengers using aircraft with more than 38 seats. All aircraft directly owned and operated by Qantas, (Australian Airlines) and Ansett fall into this category and are therefore operated at or above the minima for the highest level of safety regulation. The levels of safety regulation beneath this category are low capacity RPT (including some Qantas and Ansett subsidiaries and partners), GA charter, aerial work, private operations and down to sports aviation. Variations in minimum standards cover both operations and airworthiness. The latter includes variations in maintenance standards where there are also two levels depending on aircraft size and style of operations.

Concerns raised during the Inquiry into Aviation Safety: Commuter and General Aviation Sectors (HORSCOTCI, 1995) included the operation of aircraft with lower regulatory standards under code-sharing or partner agreements without public knowledge. Qantas liveried partners included aircraft as small as the BAe Jetstream 31 and Cessna Titan. The latter aircraft needs either one or two pilots depending upon whether it is operating charter or low capacity RPT and there was genuine concern as to whether passengers were fully aware of the difference. A ticket bought on Qantas between London and Bourke may end up with the last leg being completed aboard a charter aircraft. Former Qantas, Director of Operations, Alan Terrell considered that to be misleading. (HORSCOTCI, 1995) Following the Inquiry, both Ansett and Qantas took a more active role in ensuring standards within its partner and subsidiary operations were in line with those of the parent airline.

The differing levels of regulation and the different accident rates is no coincidence although that is not to suggest that accident rates are directly a function of regulation. Smith (1992) noted accident rates for five levels of air operations based on US and Australian accident statistics to be vastly different.

Table 10.2   Air travel safety

Type of Operation	Fatalities per 100bn passenger kms
Airlines	19
Commuterlines	374
Charters	967
Private Business Flying	10,666
Ultralights	60,000

Source: Adapted from Smith, 1992.

At the level of the major RPT carriers, the role of the regulator has always been important, not least because the level of regulated safety was that much higher than for other classes of operation. However, the importance of *self regulation* should not be forgotten in terms of airlines complying with standards, not just in case of surveillance by the regulator, but because they believed them to be proper. The relationship between the major airlines and regulator has always been close, but not at the stage of capture. Even in spite of this, there have been many times where the airlines have not been under close scrutiny, sometimes because they are ahead of the regulator. In some instances, carriers have taken their lead from the UK CAA or US FAA where regulations do not exist within Australia, only for the Australian regulator to eventually follow suit. Whilst such a process is satisfactory when the airlines are acting responsibly, it is open to some abuse and suggests that the regulator has not been able to execute its function adequately.

While safety regulation is designed to act as a framework on which the safety of a national aviation system is based, it is primarily concerned with setting minimum standards. Whilst higher minimum standards in one state compared to another may see a generally higher level of safety, it is more likely that major airlines see regulations as only the bare minimum. As such, differences between ICAO member states which accept recommendations in a near unadulterated form are likely to be quite small. The significant difference will be in the interpretation of those regulations by airlines, airfields, ATC providers and the surveillance arm of the regulator. In Australia, conservative regulation is indicative of a wider industry attitude towards safety rather than the powerhouse behind it.

11 Infrastructure and Support

Airport Integrity

Airports vary considerably in the quality of service they provide. This is not simply in terms of facilities for the passenger or cargo customer, but also in terms of the service they provide to aircraft. There are a number of influencing factors including,

- The size, surface and condition of runway and taxiway pavements
- The availability of approach aids
- Obstacle clearance
- Prevailing weather conditions
- Surrounding terrain and approach procedures
- Level of traffic vs ATS provision.

Variations in which types of aircraft use a particular airport, as well as issues of crew familiarity, combine to make assessing the operational quality and safety of particular airports quite difficult.

The International Federation of Airline Pilots' Association (IFALPA) have a black-list of airports and airspace around the world (known as Annex 19) which do not satisfy the standards of the association to one degree or another. Barlay (1991) lists sixty airports or countries where airports are renowned for being particularly poor and are included in IFALPA's list. It is not a list of third world or secondary airports as may be expected, but contains big international airports including fourteen in the USA. Even large airports such as New York JFK and La Guardia are included for their lack of safe overrun areas. None of the sixty are in Australia and Qantas only fly to six of them.

Size, Surface and Condition of Runway and Taxiway Pavements

Folklore suggests that the two most useless things in aviation are fuel left in the bowser and runway behind the aircraft. The length of runway is an issue for aircraft at both take-off and landing particularly when emergencies arise such as aborted take-offs or misjudged touchdowns. However, it is not as simple as saying that the longer the runway, the greater the safety margin it provides. To investigate whether an extra margin of safety is assured at Australian airports by aircraft using runways in excess of their operating requirements seemed to be a worthy aim.

However, the performance of aircraft at take-off is affected by a variety of factors including weight, outside air temperature and pressure, wind and the need, or otherwise, for derated or noise-abated take-offs. This makes direct comparison quite difficult. Crews derive the required take-off runs from charts provided by the aircraft manufacturer and kept on the flight deck. These specify minimum required distances for take-off based on the extremes of all the controlling factors, but this does not provide a useful guide for answering the above question.

The original method to test the question was to compare the mix of aircraft types and their minimum requirements to runway lengths. That is to say that for a sample airport such as Melbourne Tullamarine, built for long range *runway-hungry* aircraft such as the B707 or B747; what is the mix of short haul, short take-off and landing requirement types such as the later series B737, BAe 146 and propeller aircraft (e.g. Saab 340) and the long take-off and landing types. This method proved rather more difficult than originally anticipated and unable to be used reliably. Interviews with airline pilots pointed out several problems:

Aircraft that required shorter distance take-offs than the runway provides often take the opportunity to execute derated take-offs. This is not only to preserve engine components and use less fuel, but also for noise abatement reasons. Departure records do not show whether take-offs were derated or not.

Flights using the same aircraft types fly different route lengths, have different passenger and freight loadings and will vary in equipment configuration between airlines. Added to this the weight differences of the varying luggage allowances between classes and the varying construction weights of business, first and economy class seats.

Pilots claim that their flying technique will be adjusted to the length of runway available. Aircraft needing short runways landing at airports with long runways will tend to take advantage of slower deceleration for the sake of passenger comfort and brake wear. Also excessive runway length (or width) can affect visual perception and make landings more difficult and are more likely to include variations in slope.

The Availability of Approach Aids

Accidents which occur on final approach account for the greatest proportion according to Boeing (1996) who estimate 23.6% of accidents occur in this phase (3% of the flight time). Hazards include undershoot, overshoot, veeroff and both stalled and heavy landings.

Precision approaches are those which utilise instrument or microwave landing systems (ILS / MLS) whereas non-precision approaches are those which use either no aids or NDB, VOR/DME equipment. Whilst larger airports such as Sydney and Melbourne are fitted with ILS equipment, many of the smaller airports served by jet traffic are not.

Table 11.1   Australian airports with ILS available on at least one approach

Sydney Kingsford Smith	Melbourne Tullamarine
Melbourne Essendon	Perth
Adelaide	Canberra
Brisbane	Tamworth
Cairns	Darwin
Hobart	Launceston
Alice Springs	Townsville

Other airports use non-precision VOR/DME or NDB approaches. This does not necessarily mean that safety margins are lower at these airports, rather that their operations are more limited by meteorological conditions. In conditions of poor visibility, aircraft will be forced to divert to the nearest alternate airfield, an eventuality which is factored into carried fuel loading. On overseas routes, most of Qantas' destinations have at least one

ILS approach available and 60-70% of designated alternates also do.

The Flight Safety Foundation CFIT Checklist (FSF, 1994) provides a tool for assessing the risk of a controlled flight into terrain accident at any chosen airport. It assigns a value to the presence of various hazards and countermeasures. It is based on the findings of the FSF CFIT Taskforce and provides an indication of the relative value of approach aids. For example, the presence of a procedure whereby the pilot (nonflying) independently verifys minimum altitude versus DME for a VOR/DME (non precision) approach is worth +20 points. The presence of second generation GPWS or better in the aircraft is worth +30.

In contrast, the presence of an ILS approach scores zero, a VOR/DME (non-precision) approach scores -15 and 'if this is a non-precision approach with the approach slope shallower than 2.75°' scores -20. An approach using solely an NDB (also non-precision) scores -30, which is equal to the score of an airport with no ATC radar capability. The fact that non-precision approaches only deduct a limited number of points in comparison to relatively simple aircraft avionics (GPWS) or training solutions is an important consideration. In other words, the ability of approach aids to prevent accidents is significantly less than human centred training solutions. This view is supported by the data contained within Boeing's Accident Prevention Strategies (Boeing, 1993). In an analysis of 138 hull loss accidents where 606 prevention strategies were highlighted (an average of 4.39 per accident), the top three ranked strategies were *flying pilot adherence to procedure* (49%), *other operational procedural considerations* (45%) and *nonflying pilot adherence to procedure* (29.5%). *Maintaining approach path stability* is ranked eighth (15%) as an *aircrew* factor, but the *availability of approach aids* is 30th (2.5%) suggesting that other factors are more important such as adherence to procedure or response to GPWS.

On this evidence, it seems to be the case that the presence of approach aids is not the major factor in preventing approach accidents. Other aspects of the natural environment and crew performance play a more significant role. A separate investigation by the Flight Safety Foundation for the Netherlands Director General of Civil Aviation examined the influence of precision approach systems on accident risk. (FSF, 1996) Using a sample of 557 airports and 132 accidents, the team concluded that 'On a worldwide basis, there appears to be a five-fold increase in accident risk among commercial aircraft flying non-precision approaches compared with those flying precision approaches.'

Whilst such a finding would seem to support the presence of ILS / MLS as a critical safety aid, it cannot be taken entirely on face value. FSF note that when stratified by ICAO region, the risk of non-precision approaches varies from three-fold to almost eight-fold. There are also other associated factors such as the absence of charted approach procedures, absence of terminal area radar coverage and '...many factors that influence overall approach-and-landing risk are outside the direct control of the airport or authorities'.

There are obvious benefits of precision approaches, but the absence of them does not necessarily equate with a poor level of safety. Whether an airfield is equipped with ILS, and at whatever category, is a function of both its need (prevailing weather conditions) and the level of traffic (cost). The approach minima associated with each category are designed to facilitate a uniform level of operating safety for each mode.

In terms of visual guidance systems, such as VASIS, T-VASIS or PAPI; main airports in Australia tend to be fitted with T-VASIS, a more accurate and flexible version of the original VASIS, pioneered in Australia. However, the significance of such equipment on the safety record is taken to be minimal. The Flight Safety Foundation, in a study of approach and landing accidents, observed that 'Though visual approach guidance (VAG) is deemed an important landing aid, no association was demonstrated between the presence or absence of VAG and accident risk for the sample considered.' (FSF, 1996)

Obstacle Clearance

The presence of obstacles at the critical climb-out and approach phases of flight can form a latent condition in the aviation system which may later combine with active failures such as engine failures or unstable approaches. ICAO *Annex 14* (Aerodromes) details the minimum requirements for clearance surfaces around airports and these have in turn been accepted by member states including Australia. In considering differences in safety margins, consideration may be given not just to adherence to the ICAO recommended minima, but also to how far these are exceeded.

In accidents where aircraft have under-, over-shot or veered off runways, the severity of the outcome has often been dictated by exceedances of the required minima for obstruction clearance. As 80% of aircraft accidents are known to occur within 500ft of the active runway centreline and 3,000ft of the runway thresholds (Hewes and Wright, 1992), the secondary safety effect of clear surfaces, whether intentional or not, is not insignificant.

Had the M1 motorway passing East Midlands Airport not run through a steeply sided cutting, the consequences of the 1989 Kegworth air disaster may have been less serious. Similarly, the effect of other accidents such as the overrun of a USAir B737 at La Guardia Airport in 1989 and aborted take-off overrun accident in Kinshasa in 1996 may have been less severe if surfaces around the airport had been different. However, such statements are based upon speculation and proving them would be difficult. It is possible that in areas with relatively poor obstacle clearance, such as the approach to Hong Kong's old Kai Tak airport, performance on the part of the crew would be modified in anticipation. Awareness of a hazard which is perceived to be extreme is enough to heighten alertness to a point where performance is improved and the hazard adequately counter-balanced.

The lack of accidents involving jet aircraft in Australia makes it difficult to speculate on the subject of obstacle clearance in excess of ICAO minima. Suppose an aircraft is making an approach to Sydney to land on the shorter, third runway 16L which juts out into Botany Bay. The crew is likely to be well aware of the fact that landing distance is more critical and would therefore aim to touch down close to the threshold or initiate a go-around early if required. Similarly, if an aircraft was to approach the longer 16R runway at Sydney, the crew would recognise that they had a longer area to decelerate which in turn could impact upon their go-around decision making.

Prevailing Weather Conditions

As discussed earlier, meteorological factors can have a significant effect upon aircraft operations, particularly around airports where aircraft are flying at slower speeds and are more susceptible to phenomena such as microburst / windshear. Differences in climate mean that each airport will experience a unique set of weather conditions and therefore some airports lend themselves to easier use than others. Airports such as Sydney and Melbourne experience good *CAVOK* conditions for a large proportion of the time; a fact which is reflected in the level of instrument landing systems.

Sydney Airport, for example, is equipped with category I instrument landing (ILS) capability whereas airports such as London Heathrow are equipped to Category III. The primary reason is the prevalent weather conditions, but it is all but impossible to speculate as to which operation is more safe. The five categories of ILS equipment (Cat I, II, IIIa, IIIb and IIIc) require different minima in terms of runway visual range and decision height, which should all equate to similar levels of operational safety.

The importance of weather conditions is related not just to the provision of instrument landing systems at airports, it may also be linked to other factors such as runway pavement surfaces, commercial and other operational pressures. Runways which suffer from excessive rubber contamination can become prone to aquaplaning in wet conditions, commercial pressure not to miss an arrival or departure slot may encourage crews to fly in marginal conditions and the threat of *get-home-itis* at the end of long trips can also affect judgement.

Whilst it is possible to say that the *prevalent* conditions at most major Australian airports are favourable, it must be balanced against extreme weather conditions and the level of available navigational aids.

Surrounding Terrain and Approach Procedures

A Flight Safety Foundation study of accidents during approach-and-landing (FSF, 1996) concluded that 'Worldwide, the presence of high terrain around an airport did not appear to significantly increase accident risk compared to airports without terrain.' However, there is a heightened perception of risk associated with operating into airports with surrounding high terrain or complex approach procedures. Accidents such as the 1991 loss of a Pakistan International A300 on approach to Kathmandu, Nepal occurred in an area of high terrain, but the principal causal factors were not the height of the ground. The investigation revealed this incident to be primarily the result of a deflected VOR/ DME beam.

The loss of a Piper Chieftain at Young, NSW in 1993 (BASI, 1994) highlighted to the Australian industry the potential for disaster from unstabilised approaches to airports with non-precision approaches, even in areas of relatively flat terrain. The aircraft descended below minimum safe circling altitude after crossing the airport NDB and collided with terrain only 275 feet about the airfield elevation with the loss of 7 souls. This was one of the major incidents which precipitated the HORSCOTCI inquiry into aviation safety in the Commuter and General Aviation sectors. (HORSCOTCI, 1995)

There are a number of reasons which would suggest that a similar accident would not happen to a jet aircraft in Australia, such as the mandatory provision of two or more flight crew and the fitment of GPWS equipment. However, the many organisational issues that were highlighted by the BASI investigation (BASI, 1994) demonstrate that issues such as the height of surrounding terrain and meteorological conditions on the day were co-incidental to the numerous latent defects that were present within the carrier's operation. The important message is that terrain and approach procedures, as relatively fixed variables in the operating system, are of less significance that the more variable aspects of the human environment.

Changes in approach procedures, or those which are difficult to understand, especially by foreign crews who may lack local knowledge or familiarity are a different hazard. Accidents such as the loss of an American Airlines B757 near Cali in 1996, highlight the danger of collision risk following the loss of situational awareness. This is also a potential problem for Australian aircraft, especially at certain overseas ports where there is high surrounding terrain or indeed cityscapes.

Rescue and Fire Fighting Provision

Under the recommendations of ICAO *Annex 14*, chapter 9, aerodromes are assigned a category for rescue and fire fighting service (RFFS) cover based on the dimensions of the aircraft using them, as adjusted for their frequency of operation. RFFS cover may be provided by public or private organisations, located in such a way as to be able to meet the following response criteria: The operational objective is for the first RFFS tender to get from the initial call to a position where it is able to apply foam at a rate of at least 50% of the ICAO specified discharge rate within 2 minutes. The response should not exceed 3 minutes to any part of the movement area in optimum conditions of visibility and surface conditions. (ICAO, 1990)

However, the ICAO formula allowed the use of a remission factor to be applied when the number of movements of the largest aircraft is fewer than 700 in the busiest three consecutive months of operations. In this instance, the category may drop by one level. It is therefore conceivable that a large aircraft, operating infrequently into an airport, may not have adequate extinguishing agent available in the event of a fire. (Taylor, 1988)

The ICAO recommendations for RFFS only become legislation through the actions of the member states and there are a large number of airports around the world that do not have the cover which is recommended for that level of traffic.

Rescue and Fire Fighting Cover Reduction

Lamble (Fire International, 1993) reported that, in Australia, there was '...no specialised Civil Aviation Authority fire and rescue services at many of the country's 400 regional airports'. This means that aircraft as large as B737, F28 and BAe 146 are operating into airfields where the rescue and fire fighting function is covered solely by the CFA (Country Fire Authority) and often on a retained only (volunteer) basis.

George Macionis, then general manager, CAA fire and rescue services commented that CAA policy was to cover around 90% of air passengers to or above those recommended by the ICAO *Annex 14* tables. This meant that airports such as Mount Isa (QLD) and Yulara (NT) which are regularly used by aircraft up to 737-300 size (106 seats) are covered only by the local fire brigades. Even a rapid response by off-airport fire tenders is very much less effective at knocking back aircraft fires as foam application rates are, at best, one third of those achieved by airport tenders. In Queensland, a State Fire Service spokesman suggested that it would take 20 minutes for three appliances to reach Maroochydore / Sunshine Coast Airport where B737-400 aircraft land. When the British Airtours B737-200 G-BGJL caught fire on take-off from Manchester Airport in 1985, the first appliance was firefighting within 25 seconds of the aircraft coming to a stop. Nevertheless, 55 of the 137 on board were killed.

The fact that operations of this nature are within the rules of the regulator does not guarantee an absence of risk. The balance of cost and benefit - in this case the balance between the cost of RFFS cover against the likelihood of a fire occurring at one of the uncovered regional airports is one that has been judged to be *acceptable*.

Yet this situation has not always been the case in Australia. Before the Civil Aviation Authority became a Government Business Enterprise (GBE) and was required to cover its own costs through *user-pays* charges, major airports were provided with cover in excess of ICAO minimum guidelines. The current situation is compliant with the regulations and is therefore *legally safe*. This may, however, be less safe than the previous set up and is certainly less safe than ICAO intended for domestic aerodromes.

Aircraft Loading

In the busy apron area, aircraft are susceptible to damage from ground handling which is frequently not reported, either through omission or to avoid punishment. The large number of people moving around an aircraft during turnaround makes such events a concern for all aircraft operators. 'The cost of ramp accidents to the aviation industry Worldwide was US$2 billion annually, of which some US$750 million was in respect of repairs to aircraft damaged on the ground.' (Kilbride, 1996)

The loading of cargo can also cause problems if stowed incorrectly. An out-of-balance centre of gravity will affect an aircraft's flying characteristics, as can cargo which shifts during flight. The carriage of dangerous goods when incorrectly packaged, labelled or mixed can also cause in-flight problems. There are relatively few large passenger aircraft accidents caused by deficient aircraft loading, although there have been a number involving freighters. Crashes due to centre of gravity imbalance tend to more prevalent in the general aviation sector where a relatively small movement can significantly effect the stability of an aircraft.

The most common problems for large aircraft come from hazardous cargo or ground damage. Hazardous cargo may include oxidising agents, poisons, flammables, corrosives and even livestock. Although many passenger aircraft types include halon fire extinguishing equipment in belly hold areas, this is not the case with deck hold cargo on freight-only and *Combi* type aircraft. The total loss of a Valujet DC9 in 1996 in the Florida Everglades was primarily the result of an uncontrolled hold fire caused by the carriage of time-expired oxygen generators as belly hold freight. The intense fire burned through control lines and entered the cabin and resulted in the loss of control by the crew.

Ground damage has, thus far, caused few major accidents. Standard Operational Procedures require flight crew to conduct a visual walk-round inspection of the aircraft prior to departure. However, in practice, such an inspection is rarely at the point where no more ground handling equipment is to be moved around the aircraft. The majority of incidents are in the form of scrapes, indentations and perforations of the aircraft skin. (Ashford, Ndoh and Brooke, 1995) It is when such damage weakens or punctures the aircraft's pressure skin that problems can occur with the flying aircraft.

Indeed, it is damage to the pressure skin which does not cause an immediate breach that is perhaps the greatest threat to an aircraft. An aircraft that fails to pressurise will not climb above normal cabin pressure height. However an in-flight failure of the skin at altitude can result in the explosive decompression of the aircraft and possibly either the loss of structural integrity or physical damage to persons on board the aircraft through frostbite and hypoxia. Although the loss of a JAL B747 in 1985 was proven to be the result of a structural failure caused by faulty repair, it highlighted the potentially catastrophic consequences of an explosive decompression.

Whilst the level of ground damage in Australia is significant, it was not flagged a particular danger by any of the expert witnesses until pressed. This may reflect one of two things; either that the problem is indeed minimal, or that it is underestimated. Various experts pointed to the relatively quiet aprons at Australian airports and the fact that ground handling was currently the responsibility of Ansett or Qantas and not third-party contractors. As traffic increases then the threat may be expected also to increase. However, in anticipation of this, Ansett and Qantas have invested in innovative ramp teambuilding and technical training programs. There is also a move towards implementing overseas training solutions such as the SCARF (Safety Courses for Airport Ramp Functions) program operating under the European Commission. (See McDonald and Fuller, 1994 and McDonald, White, Fuller, Walsh and Ryan, 1993.)

The geographical layout of Australia dictates that a relatively high volume of freight is carried by aircraft. This is in the form of dedicated freighter aircraft and as belly hold cargo, with Qantas also using *Combi* aircraft until 1995.

BASI figures for 1990 (BTC, 1991) indicate that of 3,830 recorded incidents, fourteen involved dangerous cargo which is less than 0.4% of the total. These incidents cover all types of aircraft including general aviation. Between 1992 and 1994, of 95 reported dangerous goods incidents in Australia, only five involved declared dangerous good shipments. The remaining 90 involved hidden or undeclared dangerous goods including those carried on board by passengers.

In 1983, the Australian Dangerous Goods Air Transport Council (ADGATC) was formed from airlines, freight forwarders, travel agents, education and training institutions, airport and regulatory authorities. (Huggins, 1995) The council is one of only a handful of similar bodies around the world and lobbies ICAO and IATA on behalf of the Australian industry. For example, approved packaging for dangerous goods was competing with similarly looking, less effective alternatives. Although the United Nations operated a marking system to identify approved packaging, it was not a requirement for aircraft use until the ADGATC persuaded ICAO and IATA to amend the rules.

Whilst dangerous goods or shifting cargo accidents have been relatively few in number, there is considerable potential for catastrophe. This was highlighted by the loss of the Valujet DC9 and the crash involving an An-32 at Kinshasa, Zaire which was overloaded and failed to climb causing over 300 deaths and 253 serious injuries in 1996. Australia seems to have experienced good quality aircraft loading and ramp handling which is not a function of any specific innovation, but rather the expectations and operating culture of the organisations involved. Aided by a relatively low level of traffic and a lack of adverse weather conditions on the airport apron, this has helped to maintain a good level of ramp safety. However, there is significant potential for future incidents as traffic density increases and other third-party handling agencies enter the market.

Air Traffic Control

As soon as more than one aircraft occupies the same area of airspace, there is a need for some sort of control strategy to avoid collision. In the early days of aviation, in the absence of radio communications, the only available strategy was *see and avoid*. This type of control is reliant upon the vigilance of pilots to spot other aircraft, their ability to take evasive action, and upon weather conditions which allow clear sight. Should an aircraft enter cloud, the see and avoid principle becomes a useless form of control.

Loss of visual reference was also a problem for navigation. In the early days, especially as aviators used physical features such as roads, railway lines and rivers for routes. Indeed, in the relatively featureless terrain of Australia, navigation was particularly difficult. The use of radios for positional-vectoring and communicating with the ground brought numerous benefits after their introduction in the 1920's. However, such facilities were limited primarily to the USA and Europe and it was 1934 before the first air radio stations were built in Australia.

'Five major aircraft accidents in Australia ...had an important effect on the development of the air traffic control system.' (Bigsworth, 1988) These are discussed in greater detail in chapter five with particular respect to the introduction of radio and navigation aids. The early development of air traffic control around the world was somewhat *ad-hoc* as demands were led by geographical and industrial constraints. In Europe, flights tended to be over many small countries and consequently involved the crossing of a number of borders, whilst in USA, the home of mass production, congestion became an issue. In Australia, aircraft flew over vast areas of unpopulated country and could easily become lost.

Following the loss of Stinson VH-UHH *en route* from Brisbane to Sydney in 1937, an aircraft which was not fitted with a radio, the Amalgamated Wireless company were called upon to set up and run an expanded network of aeradio stations. Whilst the Inquiry claimed that radio would not have prevented the accident, the political impact was such that it provided the impetus needed to set up the aeradio system. Communications Officers who manned the new system worked in relatively poor conditions with the help of pilots operating in the outback, creating a spirit of co-operation that '...has continued to the present day' (Charlwood, 1981). These officers did not control traffic; all decisions were left with the pilot.

A second, major fatal accident occurred in October 1938 when DC2 *Kyeema* struck Mount Dandenong 20 miles beyond Melbourne airport in broken cloud. Evidently the pilot had been mistaken in his visual pinpointing of the ground by 20 miles and had declined the offer of a directional fix from the ground communications officer. The subsequent investigation was fiercely critical of government delays in providing pilots with direction finding equipment and considered what may be done to make civil flying safer. A system whereby movements of aircraft could be checked by a competent person on the ground was recommended. Essentially, this was the birth of air traffic control in Australia. It was followed by the recommendation that experienced airline pilots be appointed by the Department of Civil Aviation as Flight Checking Officers (FCOs).

Australia's size meant relatively few aircraft were operating over vast areas where navaids were few and far between. The pilot therefore needed protection against his own mistakes to avoid being forced down without a clear idea of his whereabouts. The FCO's role was to assist a pilot in his preflight planning and check his position during flight. This was similar to flight dispatchers elsewhere in the world, except in Australia, they also had the power to refuse approval of a flight plan if they felt it did not match the regulator's requirements. Pilots were required to estimate the length of flight and carry enough fuel to reach an alternate airfield with one hour's fuel left in reserve. Minimum cloud base and visibility were laid down for safe landing at each airport and if conditions fell below these minima, the FCO was obliged to close the airfield and demand a diversion. Position had to be passed from pilot to aeradio officer, although the FCO also had to check if it was reasonable for the pilot to be in that position. Finally, it was the FCO's task to keep the pilot informed of changing weather conditions and nearby aircraft.

The FCO was a function only provided in Australia and although they sometimes made unpopular decisions such as diverting aircraft, the whole aviation community realised the system was protected by these highly experienced officers. Most were senior airline captains with 10,000+ hours, familiar with aeronautical decision making and not to be disobeyed. Whilst this was a development well before the jet age, its significance to this case study is in terms of the mantle of safety which became *expected* from the air traffic system.

The post-war era saw a general realisation of the potential of air transport. Servicemen had come to accept air travel as something natural and there were thousands of now spare aircraft and skilled airmen who were keen to continue their aviation career. The 1944 Chicago Convention which established ICAO acknowledged the need to standardise services to aviation, although the time between the end of the war and the publishing of ICAO's recommended practices and standards was a difficult period.

FCO's returned, but to a very different system. The USAF had left runways, hangars and control towers and introduced radio telephones, allowing controllers to speak to aircraft for the first time in Australian skies. By the end of 1945, the Department of Civil Aviation, aware of the risk of mid-air collision, decided aircraft flying the same route at the same altitude should be separated by ten minutes flying time. How this was to be done was a problem eventually solved by a Sydney FCO. A simple computer made of a glass disc and clock face allowed the positions of aircraft to be calculated from its airspeed and departure time.

In 1946, the first Superintendent of Air Traffic Control was given the task of organising Australia's system of control. With help from the 1937 ICAO conference in Melbourne, the final plans were decided and a system was created which has continued to evolve up to the present day. Australia was separated into a number of Flight Information Regions (FIRs) each named after its main centre. An air traffic control unit was set up in each FIR generally augmenting an existing aeradio station. Control areas were created along major air routes and permission to enter could only be granted by ATC. All aircraft wishing to operate in controlled airspace therefore needed two-way radio.

Australia also decided to treat all RPT aircraft as if they were unable to see each other, as if they were flying in cloud. In most other countries this is not the case, where pilots flying in clear weather can rely on their ability to *see and avoid*. Whilst traffic densities were much lower than those in the US, Australia decided that the potential problem of delays caused by ATC separation was outweighed by the risk of having controlled and uncontrolled traffic in the same airspace. Therefore, strict minimum separation distances were enforced across the whole system from the beginning.

At International conferences, Australia supported its decision by emphasising the risk of deteriorating weather and the lack of decision and action time for two aircraft approaching at closing speeds up to 1,000 knots. Recognising the heightened risk where aircraft converged around aerodromes, control zones were established. Within these zones, the restricted airspace extended to the ground unlike in control areas which allow uncontrolled traffic to operate below certain altitudes. For aircraft operating outside control areas and therefore not separated by ATC, they were required to report their whereabouts to Aeradio / Flight Service who then passed this information to other pilots. Each pilot was then responsible for maintaining adequate separation.

On 2nd September 1948, DC3 *Lutana* struck terrain near Nundle, NSW with the loss of 13 souls. The aircraft was about 100 miles off course but still in radio contact. Whilst the inquiry failed to determine the cause, ATC was highly criticised. Following the 1947 ICAO conference, the *Lutana* disaster and international investigations by aviation experts, a new ATC system was shaped.

Air traffic control was split into three sections: Traffic separation became the responsibility of airport control (in the tower) and a new area control section was formed. The third area retained the FCO's task of providing individual pilots with a safety service and became Operational Control. This was the start of modern air traffic control, which although supplemented with new technology, has remained largely the same since.

The function of Operations Control, as provided by the then Civil Aviation Authority, ceased in 1992 with responsibility passing to the pilot in command or airline. Whilst this was intended to reduce the use of resources, especially as the CAA had become focused on *user pays*, the by-product was to remove a safety margin from the operating system. For example, up to 1992, Operations Controllers were able to close a runway because of inclement weather. After that date, flight crew were given full responsibility for making a decision to land in poor weather; a decision made with one less expert in the loop. Arguments from pilots that Operations Controllers were closing runways at times when the weather was still ok, because of a lack of operating experience or information may be counterbalanced by other occasions where pilot decision making is marginal for similar reasons. As the Operations Controller was only able to make a 'fail-safe' decision (i.e. he could not command a pilot *to* land, only *not* to land), the removal of his input from the process may be seen to be a reduction of the safety net, although to what effect is not immediately apparent.

The role of Flight Briefing Officers (FBOs) was also removed in the early 1990's under the controversial Review of Resources within the Civil Aviation Authority. The Briefing Officer was responsible for collating flight planning information from flightcrew for entry into the air traffic system. Whilst it was the responsibility of the flightcrew to prepare the information, the FBO also acted as a safety net to make sure all of the necessary details were correctly formatted and forwarded. Technology has made it easier to submit such information, although once again, how much effect the removal of the FBO as a safety margin has had is not readily obvious.

Collision Risk and Traffic Density

Ashford (1994) observed that for 219 large aircraft accidents, only 3.7% were considered mid-air collisions. Unfortunately even though the review was of causal factors, it does not discuss what other factors (such as lack of situational awareness or failure to look out.) were involved in each accident. It is also possible that as mid-air collisions, by definition, involve two aircraft, that the number of accidents is effectively doubled for incidents involving two (over 5,700kg MTOW) aircraft. A review of the Boeing accident database (Boeing, 1993) reveals a total of 22 mid-air collisions from a total of 962 accidents (2.3%). Of these, ten accidents were fatal (two involved two passenger aircraft). Air Traffic Control deficiency was deemed primary cause in six (three fatal), crew error in eight (four fatal) and undetermined/other accounted for the remaining eight (three fatal).

Although there are a number of recorded cases of mid-air collisions, few are deemed to be the direct result of high traffic density. However, in September 1976, a Trident 3B and DC9 collided over Yugoslavia with the loss of 176 souls. Both aircraft were under the control of Zagreb Air Traffic Control Centre which was the second busiest centre in Europe (Stewart, 1994). Controllers were working at overload using procedures that could allow just such a system failure to occur. The active controller, who had worked three consecutive 12 hours shifts and was covering the late arrival of another member of staff, became task saturated and lost track of the location of the two aircraft. The accident was not simply a function of the traffic density, but more the ability of the ATC to cope with a particular level of traffic.

Other factors which increase collision risk include the presence of aircraft in the wrong position. In 1978, a B727 collided with a Cessna 172 over San Diego after the flight crew lost sight of the Cessna and assumed they must have flown past it. The light aircraft had turned onto an unauthorised heading, which was a collision course with the B727. A third aircraft is speculated to have been in the area also without clearance and this may have confused the B727 crew as to the real position of the C172. Once again, the systemic cause of the crash was the inadequacy of the control strategy to separate aircraft (particularly when one or more aircraft is flying contrary to clearances) rather than the traffic density.

These two examples had tragic outcomes, but represent only the tip of the iceberg as far as breakdowns in separation go. In terms of system safety, a breakdown in separation may be considered to be every bit as serious as an accident. Studying statistics for near misses reveals that such occurrences are not just a function of traffic density. There are, for example, instances when particularly low traffic density can lead to low arousal and therefore poor vigilance. In May 1995, a B737 and B747 were directed onto a collision course 160 NM north of Broken Hill, Australia. A collision was narrowly avoided as a result of a TCAS (Traffic Alert and Collision Avoidance System) Resolution Advisory on board the B747.

The BASI investigation into the incident stated that there were only three aircraft under the direction of the controller at that time. However, the controller was interrupted by '...nine separate items involving 25 interchanges' (BASI, 1995) whilst attempting to perform a time of passing calculation. The high mental workload caused by the interruptions is suspected to be a reason behind the erroneous calculation that created a collision pair. High traffic density was not a factor, but controller workload was.

Ratner (1987) observes that, 'Many of the reported incidents worldwide have occurred in very low workload conditions, where inattention associated with well-known human performance difficulties in maintaining vigilance in low-stimulus environments was involved.' He also adds that such a phenomenon has also been the most difficult to correct.

When two aircraft, enter the same airspace, a single collision pair is produced. When a third aircraft is added, so the number of collision pairs is increased to three. As the number of aircraft increases, so the number of collision pairs increases at an exponential rate. Hence, the inclusion of a fourth aircraft means 6 possibilities of a collision. BASI (1991) observed that the number of collision pairs could be calculated using the following formula;

$$P = N \times (N-1) / 2$$

Where N is the number of aircraft operating in the specified area of airspace. The critical issue in considering the effect of increasing traffic density is that although the collision risk also increases, this is only the case in the operation of 'see and avoid' airspace.

See and Avoid

This most basic form of air traffic control is based upon a maritime principle for slow moving shipping (Marthinsen, 1989). It requires the vigilance and ability of flight crew to recognise and avoid possible collisions. The more aircraft in one piece of airspace, the more to spot and avoid, and therefore an increase in collision risk. (Although there is some evidence that collision risk diminishes for a while as the increase in aircraft and therefore perceived risk leads to an increased awareness and vigilance on the part of flight crew)

See and Avoid is a method of ATC that is only suited to very low traffic areas and is generally unsuitable for fast aircraft due to the closing speeds involved. Although Graham and Orr (1970) estimate that at closing speeds of between 101 and 199 knots (for example, two Cessna 150's flying head-on), 97% of collisions are prevented; above 400 knots closing speed only 47% of collisions are avoided. Two fast jet aircraft may reasonably reach head-on closing speeds of 1,200 knots where it is unlikely that either flight crew would see each other's aircraft should they be on collision course, let alone attempt any evasive manoeuvre.

Controlled Airspace

Controlled airspace describes various levels of ATC which involve an air traffic controller. These may range from procedural (non radar) control to primary and secondary radar coverage. Approaches to large airport such as Melbourne and Sydney will be controlled by a mix of primary and secondary radar, whereas small airports (where controlled) and cross country air routes (outside radar coverage) will depend upon procedural control. Radar coverage within Australia is limited as shown in Figure 11.1.

Whilst this covers the vast majority of high capacity arrivals and departures, it does leave a huge area outside of radar coverage in procedural control or outside controlled airspace. By itself, this does not equate to a degradation of safety, providing that the systems operate effectively. Procedural control, for example, demands greater separation distances than radar control to attain a similar level of efficiency. Operations outside of controlled airspace under MBZ (Mandatory Broadcast Zones) or CTAF (Common Traffic Advisory Frequency) attain an acceptable level of safety, providing that aircraft adhere to procedure.

The absence of radar coverage, does however, remove a safety net from the ATC system. Where aircraft are involved in a breakdown in separation (whether as a result of misinterpretation of control instructions or because of incorrect instructions), the controller is unable to see a possible collision pair or receive an automated warning. In the 1995 near-miss North of Broken Hill involving a B737 and B747 flying in opposite directions along a two-way air route, miscalculation of the relative position of each aircraft by the air traffic controller led to a climb clearance which brought the B737 within one second of colliding with the B747. The collision was avoided solely by the B747 crew's reaction to a TCAS resolution advisory.

At that time, both Qantas and Ansett domestic fleets carried only mode S transponders and not full systems. Had the B747 not been equipped with full TCAS, then a fatal collision would have been most likely. Had the aircraft been flying in radar coverage then the controller would have been able to see the relative position of the two aircraft.

Figure 11.1   Australian radar coverage (1999)

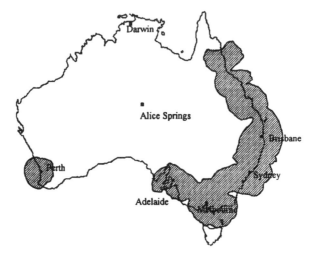

Source: Adapted from Civil Aviation Authority, 1995.

Even in areas of radar coverage, equipment failures can lead to incidents. In 1995, an Ansett B737 suffered a failure to its SSR transponder and in effect became invisible to primary radar. This went unnoticed by the controller who therefore failed to hand over the aircraft to its next sector. This situation continued for over an hour until the arrivals controller at Melbourne questioned why an Ansett aircraft was not five minutes behind the Qantas arrival as was usual. At this point, radio contact was re-established with the 'ghost' aircraft and it landed safely. The silent failure by the single SSR transponder on the B737 represented a poor level of redundancy (European aircraft carry two transponders) and was not self reporting. The failure of the controller to detect that the aircraft had disappeared from the screen also represented a grave error whereby another aircraft could have easily been assigned that altitude.

These incidents represent failures of the ATC system which are neither unique to Australia, nor directly a function of traffic density. Lack of radar coverage in the first example removed a potential line of defence, but such provision could not be justified at that level of traffic. Indeed, if traffic density were a factor, it was underload that led to controller inattention.

Traffic density in Australia is generally relatively low. Former CAA Chairman, Dick Smith (1993) suggested that at any one time, there is approximately 3% of the traffic flying over Australia as there is over the similarly sized continental USA. Indeed, there are over 200,000 aircraft registered in the US compared with less than 10,000 registered in Australia. (CAA, 1994) However, Australia's 1,100 air traffic controllers are responsible for one-ninth of the world's airspace (CAA, 1994b) and therefore the existence of a hazard cannot simply be related to the level of traffic. The possible hazards include;

1. Air Traffic Control error;
    - *incorrect or inadequate instruction / advice*
    - *failure to provide separation*
    - *failure of interface with other controllers*

2. Flight crew error;
    - *omission of action / inappropriate action*
    - *loss of situational awareness*
    - *slow / delayed action*
    - *failure in look-out (particularly in see-and-avoid)*

3. Uncontrolled events;
    - *unauthorised entry to controlled airspace / change of flight level*
    - *covert military aircraft operations.*

One of the crucial factors which should be considered is the fact that traffic is not equally spread across the Australian system. This fact is not made clear in Smith's (1993) analogy and yet is of major significance. Traffic is centred around a relatively small number of cities on the Eastern Seaboard, and Perth where traffic is often a mixture of low level private and GA operations, military and commercial RPT traffic.

The presence of severe storms, especially around Sydney, Brisbane and Perth Airports adds another dimension to the issue of traffic density. This is not least because the greatest effect of weather on aircraft is at low altitudes on departure or arrival and these are the areas where traffic density is highest. Comments from pilots regarding arrivals at Sydney included concerns about the level of traffic, complexity of arrival or departure procedures and weather conditions (such as crosswinds) being at the limits of safe operation in the name of noise sharing. Again, it is not a single causal factor that risks causing an accident, but rather a combination of several.

Review of the Air Traffic Services System of Australia

In November, 1986, Ratner Associates Inc. (RAI) was commissioned to review the Australian ATS system with regard to its adequacy for meeting the requirements of Air Navigation regulations. This represents one of the most thorough reviews undertaken of any ATS system and is therefore worth special consideration. The review (Ratner, 1987) concluded that, '...the Australian ATS (Air Traffic Services) system is basically sound. The number of reported BoS (Breakdown of Separation) incidents is less than 1 in every 50,000 aircraft movements, and the character of the incidents is similar to overseas occurrences.'

They also concluded more specifically that 'Although traffic conditions in Australian airspace often saturate the traffic-handling capacity of a sector or larger area during the busiest periods, system overload *per se*, that is, traffic levels beyond the inherent capacity of the present system, has not been a significant contributor to the occurrence of BoS incidents'. They also highlighted the fact that Australia is apparently '...relatively free of many of the problems with which other ATS systems are grappling'; mixed controlled / uncontrolled airspace, airspace intrusions, persistent occurrence of non-English phraseology and joint military / civil use of airspace with separate ATC units.

Notwithstanding these generally positive comments, RAI made a large number of observations about problems within the air traffic system. In respect of BoS, they found complacent attitudes, inadequate foresight and planning, reliance on individual technique (which may be deficient) and a lower degree of standard operational procedures (than in other similar systems). These factors were compounded by insufficient understanding of technical information regarding aircraft operation, poor supervision, a lack of standardisation and a comparatively low level of experience amongst ATC officers. This translated into a lack of motivation at all levels which has affected on-the-job training and the press for constructive change. (Ratner, 1987)

Whilst such remarks may seem damning of the system, it should be borne in mind that the review was designed to look for areas of weakness in anticipation of the reorganisation of the Department of Civil Aviation. It had already been stated in the overview that the Australian ATS system has a safety record '...amongst the world's highest', but that such '...past performance is no guarantee of future performance.' (Ratner, 1987) RAI also recognised the fact that the air traffic system had grown in complexity meaning that it was becoming increasingly stressed in certain areas. They note that 'It is necessary to rely increasingly on the backup and support capabilities of the system to ensure continued safety and efficiency. The backup and support mechanisms currently in place in the ATS system have not advanced at rates commensurate with the system's traffic loads and complexity.' Assuming that the situation was not corrected then it seems logical that increased traffic would stress the system to a point where its defences were breached and an accident occurred.

RAI revisited the Australian ATC system in 1992, at the request of the Bureau of Air Safety Investigation, in response to an increasing number of incidents. 'Comparison of incident data from the 1982-1986 period with the data from the period 1 July 1988 through 30 June 1991 indicates a higher average number of incidents in the more recent period, both on an absolute basis, and when adjusted for changes in the average traffic volume' (Ratner, 1992). The number of incidents involving large aircraft had also doubled during this period, a time which also included the restructure of Air Traffic Services from the Department of Civil Aviation to the Civil Aviation Authority Government Business Enterprise.

RAI had numerous concerns regarding the integrity of the system which varied from issues of technology deficits and airspace configuration to training issues and high level safety management. The apparent decline in standards, particularly in respect of BoS incidents involving large aircraft, was attributed to '...increased system complexity in response to industry pressures and traffic growth'. (Ratner, 1992) This situation was much as RAI had predicted in the 1987 review and this was observed to be partly as a result of '...lagged implementation of operational changes' by the new CAA management.

Recommendations for remedial action were numerous and have been addressed at various speeds over the last few years. These have included structural changes within the organisation (in addition to the CAA's devolution) such as the enhancement of the Quality Assurance role, attempts to increase human factors expertise, the foundation of a Board

Safety Committee and the development of a formal safety management strategy.

The situation remains in a state of comparative flux as major changes such as the recent introduction of TAAATS (The Aus*ralian Advanced Air Traffic System) and deferral of the *Airspace 2000* project continue. Crew fears regarding the risk of mid-air collision are high and issues of airspace management remain highly political at both industry and federal levels. The future integrity of the air traffic management system is far from clear, but the efforts towards improvement are continuing and have the potential for significant improvement.

Communications and Language

Air traffic management is no different from aircraft operation, in the respect that communication is both its strongest defence and its weakest link. FAA research conducted in 1985 (CAA, 1994c) examined operational errors and deviations in the US air traffic system and listed 98% as having human error as primary cause. In turn an CAA ATS forum held in 1994 found that '...inadequate communication is a major human factor in most failures of the system' (CAA, 1994c). A workshop conducted at the 1992 Australian Aviation Psychology Association Symposium (Hayward and Lowe, 1993) to examine ATC issues concluded that 'Effective communications between differing parts of the system in the ATS environment is essential. Indications are that in some instances the quality of this interaction may lead to a reduction in safety standards.'

In many ways, such a statement underestimates the potentially catastrophic consequences of inadequate communication, especially in areas of procedural (non radar) control where voice communications are often the sole means of separation (other than TCAS and *See and avoid*).

Cushing (1994) suggests that there are two distinct areas of communication problem which are pertinent to the ATS environment; namely those which are language based and those which are not.

Language Based Communication Problems

Although English is the recognised international language for communication in aviation, there are a number of factors which can affect the integrity of communications. When International flights enter control zones where the local language is either Spanish, Russian, Chinese or French, they may find that local aircraft talk to controllers in their native tongue. Whilst such a procedure is not inherently dangerous and controllers will be able to communicate in English as required, it reduces the margin of safety afforded by the so called *party-line effect*. This is where crew listen into the communications between ATC and other aircraft, and enhance their own situational awareness. Communications within Australian control are in English, which makes it easier for other crews to monitor communications.

In instances where English is being used as a second language (ESL), either by ATC or aircraft there are a number of problems associated with translation and interpretation. English is a somewhat ambiguous language and literal translation can be confusing. What may seem like subtleties to English speakers (for example, punctuation, intonation and inference) can cause considerable confusion amongst ESL speakers. Whilst the use of standard phraseology is designed to combat these problems, it is an ideal that often fails in times of stress, or when mixed with heavy accents.

Major disasters such as the 1977 Tenerife air disaster where a Pan Am and KLM B747 collided, and the 1976 mid air collision involving a BEA Trident 3B and Inex-Adria DC9 illustrate the consequences of a single word communication error by an ESL speaker and the tendency for controllers under intense stress to revert to their native tongue. Cushing (1994) notes an example involving a wide-body aircraft which was taxiing to the terminal and had to cross an active runway. Asking the ground controller, 'May we cross?', the reply came as 'Hold short'. The aircraft crossed the runway, narrowly missing a landing aircraft only for the subsequent investigation to find that the crew had heard 'Oh sure' and not 'Hold short'.

Strong accents are not the exclusive domain of ESL speakers and can apply just as much to regional dialects. It may be that errors made in communication between two English speakers are more likely to go unchecked than when one speaker is more obviously struggling with language than the other. In the near miss incident involving a B747 and B737 North of Broken Hill, the British B747 Captain became quite deadpan in the way he notified the controller of the TCAS resolution advisory and near miss, to the point that the controller seemed not to register the seriousness of the incident from the communication. Reaching a *controlled rage* which may be interpreted as *typically British*, the Captain communicated in a way that the controller interpreted to be 'calm and non-plussed'. An Australian or American Captain may have been more animated in the way they reported the incident.

Australian ATS and operators may be seen to have an advantage in communications by virtue of speaking English as their native tongue. It provides an additional margin of safety, although to what degree is difficult to evaluate. The temptation may be to use non-standard phraseology with other English speakers, which can then cause confusion or misinterpretation when used with ESL speakers. Conversely, a reliance solely on standard phrases does not aid understanding of abnormal situations. In an incident in 1997, an international crew landed at Sydney using standard phraseology, but could not then understand instructions for taxiing to the terminal. They were left blocking the taxiway for nearly two hours until an interpreter could be located. An ability to understand non-standard phraseology, especially in abnormal or emergency situations can only add to the flexibility of the safety system. Whilst Australian operators may operate in areas where a second aviation language is allowed, it is not to be used for International operators. Therefore, although there is a potential problem, it is theoretically limited to the presence of other aircraft in the sky.

Non-Language Based Communication Problems

Communications problems with the ATC interface are not limited to those of language, other problems are related to anything from technology to compliance. Communications technology has advanced considerably during the jet age and with the assistance of satellites, but is still susceptible to interference and failure. Loss of critical information due to poor transmission or overlapping messages has the potential for error, although standard operating procedures require such information to be verified through readback confirmation.

Problems of compliance are potentially the greatest source of communication errors. As mentioned earlier, the largest category of crew error is a deviation from standard operating procedure (33%). Whilst these are not specifically ATC deficiencies, they give an indication of a problem that affects both flight crew and air traffic controllers. Sears' (1986) analysis of 93 accidents found deficient ATC / crew communications to be present in 9% of accidents, yet this is not necessarily a fair indication of the prevalence of deviation from standard operating procedure in ATC / crew communication. RAI (Ratner, 1987) observed '...a marked tendency for ATCOs (air traffic control officers) and FSOs (flight service officers) to develop their own personal styles of operation, practices and techniques...our comparison of the Australian ATS system with others showed that this tendency is far more pronounced than in other modern ATS systems.'

In contrast, within the major carriers, the culture of strict compliance with standard operating procedures, especially within Qantas, provides a defence against the use of non-standard phraseology in ATS communications. Nevertheless, non-standard phraseology is still used, but is being addressed through additional human factors training.

Problems of compliance are not limited to the consequences of using non-standard phraseology. They also exist as a result of misinterpreted instructions or procedures which may in turn be the result of change. In the last few years a number of concerns were voiced by both pilots and controllers alike regarding the speed of change and what were seen as politically motivated revisions in approach and departure paths. This was especially the case around Sydney. A culture appeared to have evolved in the mid 1990s whereby STARs were cancelled early (where traffic permitted), to facilitate a faster arrival, and were then recommenced because of slower than expected preceding traffic. Such a practice was totally alien to the concept of standardisation which STARs were designed to embrace, and was ended swiftly in 1997 following the inaugural Airservices Safety Forum.

The relatively small Australian network and diversity of aircraft types mean that aircraft, and therefore flight crews, will operate to a limited number of ports. This can only enhance the level of familiarity associated with ATC procedures. However, long haul operations, especially for Qantas, will mean that a large number of overseas air traffic control authorities are involved and destinations may be operated to relatively infrequently by individual crewmembers. The risk is a lack of familiarity, which is addressed through an efficient, fail-safe system of chart updates and communications systems within flight operations which may include NOTAMs for information regarding approach aids.

Communications remain as important to the integrity of air traffic services as they do to pilots and there is every reason to believe that cultural strengths of open and direct communication with minimal loss of face are strengths behind the Australian ATC system. However, International destinations have always involved flying through areas of questionable communications ability, a fact reflected in Qantas' SOP that no aircraft will fly north of Darwin with an unserviceable TCAS unit.

The level and quality of support infrastructure provided for the aviation system is generally high. However, to judge the contribution of facilities such as aerodromes and air traffic control is difficult. Certain deficiencies may appear to have a very direct and clear effect on the safe operation of aircraft. However, for ICAO-compliant countries, it is all but impossible to say that airport X is safer than airport Y simply because it has, say, a higher category of instrument landing system.

Variables that can influence the safety of aerodromes include not just the level of facilities on the ground, but also air traffic control, surrounding terrain, mix of operations and so on. For example, Australia has few airports fitted with instrument landing systems, but this fact must be viewed in the context of low surrounding terrain, generally good weather conditions and good air traffic control. The human factor is also significant. Familiar aerodromes can spring fewer surprises on pilots, but may also engender overconfidence or complacency.

What seems paramount, is that facilities and infrastructure remain tailored to the operating environment they are supposed to support. A good safety record should not allow safety margins within these areas to be reduced. Innovations such as TAAATS demonstrate a long term commitment to ensuring world's best practice in ATC. However, reductions in Rescue and Firefighting that allow aircraft as large as B767 to operate into aerodromes without cover seem to go against the recommendations of ICAO and may signify a weakening attitude towards system safety.

12 Understanding the Safe System

Reaping the Case Study's Benefits

A safe system is one where hazards and countermeasures are effectively balanced in a stable equilibrium. Achieving such a balance in a complex socio-technical system such as aviation is not easy, not least because, as a transportation mode, it is a system in constant evolution. Advances in, say, technology or training will have to interact with a myriad of changes elsewhere. These may include attitude, organisational size or external influences in the form of economic or regulatory pressures. Largely opaque technical systems such as aviation make the complete understanding of all influences an unrealistic aim. Risk management techniques generally acknowledge this constraint and attempt to achieve system safety through multiple defences or filters designed to break the accident chain.

In considering the safety of Australian commercial aviation, it has been necessary to try and understand how a multiplicity of factors have interacted in a system of considerable complexity. This has included examining time-series data collected through a series of primary and secondary techniques, and covering a period where technological and human system changes have been, at times, dramatic. Aviation is still an industry in its infancy. This is illustrated by the fact that Qantas co-founder and eventual Chairman, Hudson Fysh, saw the airline's equipment go from the first *rag, stick and wire* Avro 504K to the introduction of the turbojet Boeing 707; all whilst he was still an employee. The changes that have occurred even within the *jet-age* since the introduction of the Comet and B707 have been spectacular and not without considerable impact on aviation safety. Yet in spite of the ongoing change, aircraft accident rates have remained relatively static since the early 1970s. In Australia this has meant no lives have ever been lost in a commercial jet aircraft accident.

Attempting to learn from the successful aspects of the Australia aviation system requires focus at both the micro and macro levels. Different safety issues achieve disproportionate levels of attention; either in response to accidents or indeed because of an apparent lack of accidents. It is easy to be misled as to the individual influence of different factors, and case studies have to be wary of uneven coverage. Notwithstanding this, the relative importance of different elements of a safety system can change over time to reflect advances in technology or training. For example, whilst weather remains an important influence on aviation, technological advances such as instrument landing systems, jet engines and weather radar have altered the significance of the threat.

As such, it is all but impossible to develop a universal model to describe a safe aviation system based solely upon the case study of Australia. It is nevertheless possible to highlight the relative merit of each of the areas examined within this case study based upon their effect on the accident record, perception and future potential. Analysts wishing to draw maximum benefit from this case study can do so by examining the individual effect of each component as well as its effect on the operating system.

The Risk Exposure of the System

Hazards to the safety of the Australian commercial aviation are summarised as follows.

Natural Environment

In the natural environment, anecdotal focus tended to be predominantly aimed at the positive aspects of the 'relatively benign climate' and 'good aviation weather'. However, there are a number of significant threats which deserve highlighting.

Australian aircraft are likely to encounter strong microburst and windshear conditions, especially in the tropical and subtropical regions of the country. Whilst such a threat may be seen to be obvious in the far north ports such as Darwin, it also affects the other large ports including Perth and, in particular, Sydney. Weather conditions at the latter are known to deteriorate quickly, a fact which is related to the airport's location at Botany Bay. Fear of microburst / windshear encounters was significant amongst Australian crews, largely because when it is encountered at low level, it is very difficult to recover. It is also a phenomenon which is poorly predicted and lasts for only a short time. An approach or take-off in such conditions is also prone to additional risk due to aircraft operating at maximum take-off weights or at the end of long sectors, or lack of familiarity from overseas carriers. This can be further exacerbated by commercial pressure to *get-*

in to this busy airport, approach and departure procedures which are confusing and frequently changed, and the expectation that '...it doesn't happen here'. While the number of windshear / microburst related accidents recorded in Australia is low, this does not mean that the incidence of the phenomena is insignificant. This is a threat that will grow with complacency or increases in traffic unless due consideration is given either to the magnitude of the threat or the prevention strategies currently available such as predictive windshear radar.

Other meteorological threats that exist in Australia include severe rain, turbulence, in-flight icing and dust storms. Their frequency may be lower than in other countries, which may minimise the hazard, but once again, a lack of incidence can also lead to either a complacency that there is no significant threat, or problems that stem from a lack of operational experience.

Both Qantas and Ansett are involved in international operations which means that by definition, half of their landings and take-offs are at overseas ports. These include aerodromes in all climates. Once again, long distance flights may mean that conditions are experienced at take-off in a fully laden aircraft or at the end of an long sector where the performance of the aircraft or aircrew may be degraded. Heavy crewing and flight time limitations may mean that pilots are only manually landing aircraft once or twice in a month. The potential is for extreme weather conditions, an unfamiliar crew with relatively low currency and a heavy aircraft. Whilst such a threat may be counteracted to a degree by the extra alertness level afforded by unfamiliarity, it should be recognised as a threat which can be counteracted through training.

Although flat terrain is repeatedly cited as a positive attribute of the Australian environment, it does not mean that relief around airports is insignificant. Controlled Flight into Terrain (CFIT) accidents often occur around airports and not necessarily because of high relief. Terrain around Australian airports is also responsible for weather conditions such as low level rotary shear and is supplemented by microclimatic phenomena at airports such as Sydney.

The so called *tyranny of distance* experienced both in terms of interstate and international operations has, historically, been a prime motivator for demanding reliability. Alternate landing sites in remote areas are rare and early pioneers often fell foul of a lack of ground support. Had the potential for aviation during the early 20th Century not been realised then

Australia would have been left behind by its traditional trading partners because of the huge distances involved and lack of alternative transportation networks. Even now the importance of air travel for international trade to and from Australia is huge. Add to this the practical issues of aircraft operating long sectors at high load factors and it is clear that modern aircraft are pushed to the limits of their operating envelopes.

Human Environment

The more recent trend in aviation safety seems to have been in highlighting the fallibility of the human environment. Indeed, there is human error to be found in every accident. Human error is inherent in aviation as it is in any system involving human input. Whilst understanding of human frailties at both physiological and psychological levels has advanced significantly, it remains a relatively soft and often controversial area of science. For many, it represents the final frontier in conquering safety issues, yet it remains poorly understood and the source of many heated arguments.

The traditional approach to human fallibility has been one of removing the *problem* or blameworthy individual; a fact which is reflected in countless accident investigations. Yet, the same mistakes are then repeated around the world, often with catastrophic consequences. This situation seems to have been allowed to continue because of the relatively good safety record enjoyed by aviation. In the event of a fatal crash, it is usually the fact that flight crew are killed first which has allowed pilot error to become a convenient way of closing cases of human error. The more recent trend towards understanding multiple causal factors in terms of latent conditions and active failures have begun to challenge the traditional approach. No longer can the vagaries of human performance be placed in the 'too hard basket' or blamed on errant personalities.

At a conceptual level, the essence of human error is the process of risk perception and risk taking. Decision making, which is at the heart of human behaviour is based upon assessing hazards and countermeasures to achieve an outcome which is deemed to be of acceptable risk. Individuals rarely make risk taking decisions that they believe will have an outcome with an unacceptable cost (whether it be financial or physical). Instead they perceive an outcome where benefits outweigh costs. However, the process is often affected by incomplete knowledge of the nature of the risks or the effectiveness of countermeasures. Risk taking decisions are not always obvious as being such; one of the reasons why some risks are underestimated. For example, even hesitating over a decision is a risk taking decision process in itself.

Decision making occurs not just at the level of individuals, but right through to organisational levels where corporate culture can strongly impact upon risk taking decisions. Deficient organisational cultures will negatively influence the decision making skills of its employees. Similarly, leadership which focuses on short term objectives will send signals to the line workers. Organisational behaviour becomes self perpetuating and entire safety cultures can be damaged to the point of an accident. Examples from within the Australian aviation system of poor safety cultures, especially in the GA and Charter sectors are plentiful and provide an early warning of a potential attitude shift within the entire industry culture if not checked.

Another, more specific corporate hazard is the process of change; something which is particularly prevalent in a dynamic, high technology industry such as aviation. The Australian aviation industry has undergone significant structural change, especially during the last ten years. The regulator has changed both its identity and the way its operating income is funded, the *two-airline policy* of domestic regulation ended, Qantas absorbed Australian Airlines before being publicly floated and Ansett became an international carrier. Whilst change in itself is not inherently dangerous, it does have the potential to result in operator unfamiliarity and unforeseen hazards.

A significant aspect of change within the Australian aviation industry has been the role of government. Whilst the ending of Domestic *two-airline policy* regulation was aimed at opening the system to free market forces, the playing field for new carriers was far from even. Both Ansett and Australian Airlines had massive infrastructure bases, owning not only the equipment directly associated with airline operations, but also a network of support services including travel agents, airport terminals and holiday resorts. The early attempts of new carriers (Compass MKI and MKII) failed, but at the same time serious denting the profitability of the existing carriers. The merger of Qantas and Australian airlines created a very different operation, forcing Ansett to establish international partnerships and start overseas operations. Meanwhile, changes within the CAA were precipitated by a high level of political intervention, especially following an apparent decline in commuterline safety. The creation of a Government Business Enterprise that was responsible for making its own revenue and maintaining safety surveillance was fundamentally flawed and led to the creation of Airservices Australia and CASA. Even following this restructuring, the former has been placed under considerable political pressure on the subjects of Noise Pollution and *Airspace 2000* and the latter on the subject of Board membership. Turbulent times within the industry, further hampered by party political agendas have created the risk of both individuals and corporations missing the most important safety issues.

An unusual hazard which seems indicative of current attitude is that of luck belief. Frequently cited in anecdotal evidence from expert witnesses and the perceptions survey, the explanation of good safety being a function of luck raises two concerns. Firstly, such a belief indicates a lack of understanding of the complex safety system that has worked well so far; secondly a genuine belief in a mystic force beyond human control. A poor understanding of systemic safety can lead to ill advised changes whilst belief in a force beyond human control may lead either to complacency (belief in good luck) or a form of fatalism (belief in bad luck).

Operational Environment

Within the operational environment, there are numerous technological hazards. Whilst in the strict sense, these may all ultimately be the result of human deficiencies, they are highlighted as failure sources which are specific to certain types of operation within the aviation environment. In terms of the aircraft which are operated, there is always a threat of unforeseen problems which may be overlooked because of system complexity or opacity. These include deficiencies associated with specific types of aircraft such as the DC10 rear cargo door or more general problems associated with major leaps in technology such as the introduction of jet and fly-by-wire aircraft. The Australian environment is not immune to such problems although their effects have been minimised by actions such as conservatism in type selection, as was the case with the DH Comet.

Some of the other technological problems, which may not be specific to particular aircraft types, stem from the condition of the aircraft. The continued airworthiness of any design is highly dependent on maintenance and the quality control which regulates it. Whilst many maintenance failures are not catastrophic, they can induce compounding errors by flight crew. The lower the number of failures, the less the need to rely upon the last line of defence (flight crew). The quality of maintenance is controlled both internally, through quality assurance, and externally, through the regulator. Poor surveillance by either party can combine with a lack of vigilance by the other with potentially disastrous consequences.

The quality of aerodrome facilities can also be critical, especially when aircraft are operating near the limits of their performance envelope. In Australia, an example is RPT aircraft operating at regional airports with meagre facilities. Overseas, another example is Australian aircraft operating at airports in marginal conditions, when heavy or at the end of long sectors. Other hazards around the aerodrome include ramp accidents and those involving incorrect loading of aircraft. This may be a function of ground staff not directly employed by the airlines or involved in the servicing of a particular aircraft.

Another external agency which is part of ensuring the safety of the aviation system is the air traffic control provider. In the course of an international flight, an aircraft may pass through numerous control providers of varying standards. The risk of mid-air collision or impact with high terrain is especially significant as the results are generally catastrophic. Complicated or poorly understood approach procedures, especially at airports without precision approaches (most of the domestic aerodromes other than capital cities) are a particular hazard when combined with other human factors considerations. For example, communication problems, especially associated with different cultures and native tongues are a major threat to any international operation.

The Risk Countermeasures of the System

Risk countermeasures which exist to maintain a good record for airline safety occur both by design and by virtue of the systems that aviation operates within. In other words, countermeasures may be a function of the natural environment; where Australian aviation operates, the human designed operating environment, or the wider human environment in which that works.

Natural Environment

The natural environment represents a relatively stable variable, which allows safety systems to be constructed around it. Weather is mostly predictable within climatic zones and physical geography is generally unchanging, except within the built environment. Anecdotal evidence points to 'good aviation weather' in terms of generally stable flying conditions and lack of ground icing. Certainly operations do not experience some of the extremes of North America and Europe, affording an extra risk counter-measure in normal operations. Even when weather is good in Australia, the large RPT carriers use weather radar and extra facilities of the Bureau of Meteorology through subsidy. Additional factors include a high collective level of experience which is afforded by the major airlines and the relatively small route network. Also, a lower level of commercial pressure may reasonably be expected to assert less pressure upon flight crews to operate in marginal conditions.

The relatively flat terrain reduces the need for 'hot and high' performance critical operations and those involving complex approaches or departures over inhospitable terrain. It also reduces the need to fly through icing levels and has an impact upon general weather conditions.

The *tyranny of distance* also had a significant positive impact upon the early history of aviation. Hazards associated with the distances and lack of settlement demanded solutions which led to the establishment of an industry level culture of reliability and innovation. Whilst advances in technology have changed the operating environment, there are many aspects of the original culture that have remained as expectations and common practice. For example, cross-checking by Second Officers even when not specifically instructed by SOPs.

Human Environment

Whilst the role of human designers, decision makers and operators are most often mentioned with regard to error and the fallibility of socio-technical systems, the human factor also represents the strongest element. Any successful defence against system failure has a human component associated with it. This is primarily a function of the unique ability of humans to evaluate consequences and rationalise in decision making, exceeding any computational power currently available. Although human strengths and weaknesses can be seen to vary between individuals, common elements may be highlighted as cultural traits. Ranging from workgroup and organisational levels to professional, industry and national levels, culture is the biggest single influence on human behaviour and therefore one of the most powerful influences on risk taking decisions.

Cultural strengths highlighted at national level in Australia include a high degree of individualism and a shallow authority gradient. In turn this has engendered a frank and open style of communication at both micro and macro levels. This makes cross-checking easier and facilitates the more efficient exchange of safety information. At an industry level, this is complemented by a culture of strict adherence to standard operational procedures, a fact which has often been overlooked in previous studies of the Australian culture. There is a great historical pride in the Australian aviation industry, supported by a *pioneer spirit* which has been forced by the geographically disparate location.

The culture of the aviation industry is sustained partly through the expectations of the general public and government. Safe operations have become an expectation in a country that is highly dependent upon aviation for interstate and international travel. Such expectations have a secondary effect on the allocation of resources and the priority of safety on the political agenda. Problems in the mid 1990s at commuterline level attracted a disproportionate level of attention which have assured a high level reaction to regain a level of acceptable risk by Australian standards.

At a corporate and industry level, one of the key strategies in ensuring that decision making is enhanced is training; something which exists at both a structured academic level and a less formal *lead by example* level. Airlines have long placed strong emphasis on the integrity of training for all disciplines. Formal education is supplemented by a generally high level of experience and the strength of corporate culture in setting, communicating and enforcing standards. Communications systems and styles which facilitate this process have also allowed easier introduction of industry-standard safety strategies such as Crew Resource Management, Flight Data Analysis and Ground Proximity Warning Systems.

The structure of Australia's aviation industry is very different from that of other nations of a similar size. In an area of land approximately the same as continental USA resides a small fraction of the equivalent population. There were two major domestic carriers and one international carrier with Qantas now being a combination of both. This situation was largely because of Government policy which limited competition on the domestic network to two airlines and made Qantas an entirely international carrier. As such, competition was severely limited which arguably led to a high level of economic stability. This allowed safety minded operators to exist without the pressures of lower quality predatory competitors, a situation quite different from the deregulated US industry.

The deregulation of the Australian domestic market in the early 1990s did not see any long-term competitors appear on the scene until the arrival of Virgin Blue and Impulse in 2000. The absorption of Australian Airlines into Qantas and subsequent public floatation has seen pressures both overseas and especially on Ansett Australia. A lack of experience with international operations and the rush to launch long-haul B747 operations were two of the latent conditions in Ansett behind the VH-INH accident at Sydney in 1994.

However, the ability of the industry to learn from mistakes, particularly those made by others is another important facet of industry culture. Historically, Australia was used to looking towards "the mother country" for a lead and then improving upon those ideas. There is no loss of face experienced within Australian carriers looking at what others do in an attempt to assure best practice. Expertise is drawn from throughout the world to supplement home-grown skills.

Operational Environment

In terms of issues specific to the operational environment, strengths are numerous and centre around the medium of the aviation system; the aircraft. Whilst the number of large commercial jet aircraft manufacturers around the world is small and therefore companies such as Boeing or Airbus supply aircraft to most large airlines, there are some important differences in design. A long-standing demand for reliability meant that aircraft selection and specification was extremely cautious. Aircraft with checkered histories such as the Comet were rejected in favour of proven designs. Even apparently successful designs often required significant modification to be accepted by both the regulator and operators.

Maintenance has always been a high profile area within the Australian system, not least because of the background of its pioneers and again, the critical need for reliability. The quality of maintenance within Australia is recognised globally. It is supported by the cultural factors mentioned above and comprehensive training. Buyer Furnished Equipment on aircraft supplements the margins of safety afforded by the basic design and includes items such as GPWS, TCAS and QARs.

Regulation within the Australian aviation system has traditionally been strong yet conservative, although recent crises of confidence appear to have rocked the CAA and then CASA to their collective cores. Historically, the regulator was an organisation with a great deal of operational expertise, which was therefore able to work with the industry better. A more recent shift towards employing career civil servants and making politically motivated appointments has raised many questions about the ability of the regulator to effectively do its job. However, whilst this is of grave concern in the lower levels of the industry such as General Aviation and Commuterline operations, it appears not to have damaged the airlines which have evolved into largely self-regulating bodies by following JAA and FAA best practice and assisting CASA in setting standards.

Aerodrome quality was a difficult area to assess as the effect of variables such as runway length, approach aid provision and ground facilities are directly related to other factors such as prevailing weather conditions and aircraft equipment. Further, differences in regulatory requirements are designed to ensure equivalent levels of safety in operation, regardless of, for example, category of ILS. Australian airports are maintained to ICAO standards and while many airports do not have precision approach aids, this is generally balanced by other factors such as weather and crew training.

The Australian aviation system remains generally less busy than its European or US counterparts. Airports generally have only two ground handling organisations, both owned by the two major airlines. As a result, the ramp remains less chaotic and, thus far, a controllable hazard. However, this is a situation that may change with increased traffic and as third-party ground-handling agencies are introduced.

Rescue and Fire Fighting cover represents a secondary safety measure which has rarely been called upon because of a lack of incidents. However, at major airports in Australia cover is above ICAO minima and benefits from being tasked solely with its core function, unlike many such organisations overseas. This provision seems indicative of a culture that is not complacent about secondary safety in the light of good primary safety. This provides an additional margin of safety which may one day make a critical difference to the outcome of an incident. However, it is a situation that is already changing. Rescue and Firefighting has been absent from a number of secondary airports for years. With traffic growth this had led to a situation where domestic B767 aircraft are regularly operating into aerodromes with no ARFF cover and no local fire service.

The level of ATC cover at major ports is good and Australia recently successfully introduced the most advanced ATC system in the world,

TAAATS. Concerns regarding the safety of the air traffic management system, which were raised in the late 1980s and early 1990s, are being addressed through initiatives such as TAAATS and a Board-led commitment to improving system safety.

System reliability in general has been very high with the operational and human environments not taking the hazards of the natural environment for granted. Positive cultural factors associated with national culture and historical challenges have helped to build a strong safety system with numerous safety margins. Open and frank communications have acted as the conduit by which a sound level of system safety health has developed. Herein lies the key to a successful future.

The industry needs to recognise its strengths and build on them, whilst other industries or countries try to emulate them, within their own system constraints. Observations made within this book aim to be part of the ongoing process and not evidence to be used in a reckless or critical manner. Australia's safety record has not been the result of luck, rather the outcome of a complex, but generally well designed operating system.

Future Threats to the System's Safety Health

Although the aim of this book has been to concentrate on what Australia does right, it is not intended to give the impression that all is perfect. Safety is not a utopian state which can be reached, it is a continuing battle against ever changing threats. Those who believe they have achieved a state of *being safe* risk falling foul of complacency.

There are a number of threats to the future of aviation safety which have become apparent during the course of this research. The threats perceived by flight crew and air traffic controllers are summarised as follows.

Economic pressure on the aviation industry is always high because of its intense capital utilisation and dependence on blue chip and service industries. In Australia's case, this has been heightened by significant structural changes undergone by the aviation industry within the last decade.

Cost cutting within both major airlines has been significant and placed pressure on all aspects of the operation. Whilst many fear that this will directly impact upon safety, some signs point to a more integrated approach towards both safety and efficiency. Ansett's current business recovery strategy puts safety visibly first and includes human factors training as a core methodology to achieving efficiency and safety by working smarter, not harder.

An increase in traffic, which is predicted at approximately 7%, will bring additional pressures to the industry. This will especially be the case in its attempts to compete with the rest of the Asia Pacific region, where growth is predicted to reach up to 11%. One of the challenges will be increasing pressures on infrastructure, especially air traffic services, airports, training and maintenance. Growing pains do not have to mean a degradation of safety, but the industry must adopt sound risk assessment methodologies if it is to adequately manage the change process. Problems which may have been underestimated because of a lack of incidents may become tomorrow's accidents if not catered for. For example, aircraft making approaches or taking off in marginal weather conditions where in less busy conditions this may have been avoidable.

Culture has been shown to have played a pivotal role in the safety health of aviation and changes at any level can have an indirect effect on the future. Changes at national level may be quite slow, but in a fast developing, underpopulated country such as Australia the effect of change should not be underestimated. Political motivations and public expectations have a significant effect on resource allocation and priorities. Issues such as noise pollution, particularly with reference to Sydney airport will put major pressure on operating efficiency.

Industry and corporate culture can change more quickly than national culture as events within the CAA / CASA during the 1990's have shown. Individuals, when placed in key positions, can have a significant effect on the way organisations operate. Positive examples include charismatic leaders like Herb Kelleher (Southwest) and Richard Branson (Virgin Atlantic), who have managed to assert their individual style on their organisation's operations. Negative examples are plentiful and all industries need to be aware that although accident investigation has moved on from apportioning blame to individuals, it has also shown the root of many accidents to be high level decision making at CEO or Board level.

The mid 1990's also brought changes to the way the Australian industry is regulated. A widescale regulatory review commenced in 1997, aiming to review the entire Australian regulations and to move towards harmonisation with the FAA or JAA. However, the temptation to harmonise or follow the lead, particularly of the FAA, should, at least, pay respect to the need for cultural suitability. Regulations which may appear excessive by US or European standards may be responsible for additional safety margins and any changes should be thoroughly examined for their systemic effect.

Finally, there is the constant and chronic threat of complacency, which is associated with any operating system that appears to be working. A lack of accidents is only a very rough guide of system safety health and in aviation, where the consequences are potentially catastrophic, not a sensible measure of current performance. The determining factors as to whether Australia manages to keep its clear record for fatal RPT aircraft accidents do not include the past accident record. Constant evolution of the aviation environment will require adaptation of the many risk countermeasures if safety is to be maintained.

Complacency is a significant threat, not least because it can strike at the core of the strongest aspect of the Australian system, namely the human environment and in particular, the various levels of culture which have held it together successfully so far. Whilst complacency represents a mood that *nothing needs to be done*, a further threat is from a belief that *nothing can be done*; a sort of fatalism towards factors beyond human control. The latter has been expressed by a number of witnesses as a feeling that 'Australia is due for a crash' or that the good record 'has to come to an end'. Accidents do not occur because statistics say they should and whilst a good accident record is no guarantee of future success, neither is it a bad omen.

It is hoped that this case study has revealed the hitherto underestimated human contribution to system safety and placed the more recognised environmental factors in context. In understanding how a safety system has evolved and operates, it has set out to reduce the opacity of the Australian aviation safety system. In doing so this provides opportunity for Australians to recognise their strengths and weaknesses and continue to develop a safety system which is responsive to change. The case study approach also provides an opportunity for those outside of the system to draw from its strengths and heed the warnings of the Australian experience.

Future Research

Good research seems to end up asking almost as many questions as it answers and it has always been the intention of this book to provoke thought on the subject of safety. One of the most important objectives of case study research , which looks at the entire system rather than a single focussed issue, is to gain a broad perspective of how several issues interrelate. This is necessary, not only to promote a wider understanding of how the system operates, but also to guard against focusing too much on one particular factor.

For example, the emphasis upon CRM runs the risk of replacing other prevention strategies rather than adding to them. The wide acceptance of CRM appears to be a good thing and yet there are still no standard metrics available for its assessment. Although introduced as a non-jeopardy activity, there will come a point where if CRM is that crucial to a pilot's airmanship, then it must also be assessed. At the moment, its development is driven primarily by anecdotal evidence and a belief that because the accident record demonstrates that in many cases crew co-ordination is poor immediately prior to accidents, a training strategy that aids effective communication for problem solving must be a good thing. The continuing risk is that the few individuals who sit at the back of CRM sessions and disregard the training content are perhaps the greatest hazards.

The CRM movement has risked losing focus - certain countries or cultures that seem resistant to the principles of CRM may already be safer by virtue of stricter adherence to standard operational procedures. Although both skills cover different areas - strict SOP adherence works well for normal operations and CRM comes to the fore in abnormal situations, they may both have a similar effect on the overall safety picture. This may also mean that prudent application of both skills may have an increased effect. Only a view of the big picture, allows these observations to be made.

Culture at its Many Levels

Although it was never the original intention, the subject of culture has emerged as an important theme throughout this research. The multi-level influences of various cultures from the professional, through organisational and industry, to national cultures have all been important and worth deeper examination. This is particularly important because aviation is a very multicultural industry, which is becoming ever more so. The birth of global carriers such as British Airways, which sees a number of different operators flying the same corporate flag through various franchising and ownership deals, has already led to a mixing of cultures. The development of global airline alliances only exacerbates the potential problem further. A number of cultural challenges have become apparent.

The effect of nationality on safety   As suggested earlier, the mere mention of attempting to link nationality to safety records is enough to cause political consternation, especially when done in a way that implies, 'we are better than you!' This has been one of the reasons why previous attempts have been so controversial, especially as they have often originated from the USA - a not especially modest culture. (Beaty (1995) notes that the motto of the US F-111 squadron based in the UK was *We Are The Greatest!* which was in contrast to the RAF's *Piece of Cake* attitude. He suggests that this is some sort of '...Freudian defence mechanism' whereby the '...USAF elevate the man and the RAF denigrate the dangerous operation.')

Close examination of the accident record reveals the unquestionable fact that some airlines are safer than others and some countries are safer to fly in. A deep investigation into the reasons behind this is long overdue and will ideally require the efforts of an international collaboration. This is not just to avoid the natural bias of looking at any cultural difference from the researcher's own host culture, but also to attain the integrity required for data collection to be valid and for the results to be trusted. If existing knowledge of cultural factors is correct then cultures where saving face is important will be likely to be selective in terms of the safety information they are likely to present. The argument may be that safety information should not be presented too freely to outsiders and if this is the case then safety differences become more about the ability to market convincingly. This does not help to develop a real understanding of the big picture.

The effect of organisational culture on safety   To date, studies of organisational culture tend to be particularly focused on efficiency. However, while safety may be judged to be inherently linked to efficiency, deep investigations into the effect of organisational culture on safety have limited.

What studies do exist have tended to be undertaken into the *unsafety* of an organisation. This may be in the form of an official accident report or a public inquiry such as the Royal Commission into the 1979 Mount Erebus Disaster. As a result of these studies being retrospective and usually after significant loss of life, the facts are often changed so as to minimise blame or liability for those who are left behind. People are naturally less inclined to speak about what they did wrong compared to what they did right. In the case of the Erebus Royal Commission, Justice Peter Mahon observed that he considered that the evidence produced before him by the airline management was '...nothing short of an orchestrated litany of lies'.

As was mentioned earlier, it was James Reason (1993) who stated 'Should we not be studying what makes organisations relatively safe rather than focussing upon their moments of unsafety? Would it not be a good idea to identify the safest carrier, the most reliable maintainer and the best ATC system and then try to find out what makes them good and whether or not these ingredients could be bottled and handed on?'

The effect of multiculturalism on safety   Although, by their very definition, international carriers have been operating in multicultural environments since their inception, aviation is becoming an ever more mixed industry. Already, airlines such as Singapore, Emirates and Malaysian have between 30 and 40 different nationalities within their respective flight crew divisions. Work such as the NASA/UT crew research project has helped develop an understanding of the need to tailor training across national boundaries, particularly in the context of CRM training. However, the challenge that has generally remained in the *too-hard* basket has been that of multicultural training. For example, CRM with a number of different cultures together.

A Qantas maintenance supervisor explained that although the majority of his staff were 2nd+ generation Australian, spoke with English as a first language and were brought up with Australian attitudes and values, there was a minority of expatriate staff too. Recruitment of Vietnamese engineers proved that they were incredibly hard workers although there was a tendency for them to close ranks and keep quiet to cover up any mistakes. Some problems arose when they spoke in their native language to each other, as the Australians sometimes thought they were talking about them.

It seems apparent that the training needs of the majority group were different from the rest. If they were both put through separate CRM courses then there is a risk that one or other of the groups would feel alienated and the closer integration, which is one of the aims of the CRM training, would not be achieved. What was needed was a single training programme that was sympathetic to both sets of needs without appearing condescending to either party. The answer does not appear to be out there, either in academia or industry and this is why future work is proposed.

Captain Surrendra Ratwatte (1997) of Emirates suggests that although the airline flies with 32 different nationalities of crew member, there is no major cultural problem. He purports that because there is no one dominant culture, and because individuals are aware of the fact that there are many different cultures working together, their behaviour is modified to take account of this. In fact, this process also works at the recruitment stage where applicants who do not wish to mix with other cultures will either not apply or be rejected on application. It seems more likely that problems will arise when there are only one or two cultures in minority to a main culture.

Another challenge is related to communications between flight deck and ATC and the effect of cultural variation on this process. Although most of the work in cross-cultural communication has been slanted towards flight deck, aircraft or maintenance communications, they all represent areas where crews are together for a period of time and are able to use body language (which can account for between 55 and 70% of communication). The virtual crew that is formed when an aircraft passes into a particular control sector may only last a matter of minutes before a new *crew* is formed. In the case of the aircraft this may be as they pass over different control sectors in one country, from military to civil control or different countries. In the case of ATC, this may be a changing mix of nationalities, different airlines, military and civil traffic.

Although all communications are expected to be completed using standard English phraseology, it is becoming increasingly obvious that there are a significant number of occasions when this protocol is not enough. Operational requirements can further complicate this process - one airline which flies to Australia uses a Captain, First Officer, Flight Engineer, Navigator and Interpreter, where only the latter has a working knowledge of English and is also the only non-aviator on the crew. The communication loop is slow and non-standard instructions (such as the cancellation of a standard arrival route) can be misinterpreted or filtered by the interpreter so that the message never reaches the Captain (e.g. for fear of loss of face).

The effect of economic change on organisational culture Aviation is a capital intensive industry, that is particularly prone to economic cycles. Major airlines such as Pan Am, Braniff and TWA have all been through bankruptcy protection, with only the latter surviving the economic fallout of US deregulation and the stock market crash of the late 1980's. Deregulation of the US and Australian aviation industries along with liberalisation in Europe have created new pressures, not just for the airlines, but also the regulators, ATC systems and service providers such as maintenance facilities. Airlines have also tended to change from state to private ownership, with a new urgency to make a profit for shareholders.

Although none of these changes *per se* could easily be proven to have a direct negative effect on safety standards, there is always a risk that organisational change will allow this to happen. As this case study has shown, some of these changes in management style or operating priorities have led to a degradation of safety margins either as a result of naïvety or in the name of rationalisation, where a lower level of acceptable safety is allowed.

What is relatively unknown is how these organisational changes are affected by the role of economics. As accidents are extremely expensive, theory would suggest that safety could be achieved through economic forces. Yet safety is often priced as a cost rather than a saving by virtue of it being negatively reported. The process of culture change can be relatively invisible and confused with other changes which naturally occur over time for example, as a result of technological change. The development of a conceptual tool for assessing the organisational culture of a particular firm at a particular point in time would provide a useful *snapshot* which could be compared with another assessment at a later date. The understanding of the importance and mechanics of organisational culture will need to be enhanced before the effect of external forces such as economic change can be properly understood.

The effect of technological advances on organisational culture   As technological change occurs over time, its influence on organisational culture is relatively difficult to isolate. Advances at the *sharp-end* of aviation have brought safety benefits through systems such as GPWS and TCAS, albeit with the threat of other problems such as the deskilling of crews. These have occurred alongside *blunt-end* technological advances such as the paperless office and improved communications. One important question raised by such advances relates to the relative improvement of accident rate. Although aircraft have become more advanced to minimise human error, the accident rate has remained static since the early 1970's. Are advances in technology being balanced out through some sort of systemic homeostasis?

Risk

The effect of culture on risk perception and acceptability   The historical emphasis on risk has tended to be directed from the more tangibles of probabilistic risk assessment or financial planning. Both of these areas rely on stable mathematical principles to predict failure rates in performance and as such have tended to steer away from the vagaries of human performance. In other words, the dependence on probabilities for quantifying and assessing risk does not lend itself to including the risk profile of humans.

This is partly due to the subjectivity of individuals' risk perceptions and acceptabilities and the inherent capability of all humans to both over- and underestimate risks. Nevertheless, it is the case that groups of people do display similar traits on risk taking and closer study of the influence of these groups appears to be warranted.

The 'collective programming of the mind' (Hofstede, 1980) which is culture, is in many ways a self perpetuating phenomenon. Those who group together tend to do so because they have something in common with the other members in the group. It is also true that the behaviour of members within a group is heavily influenced by culture of the group.

The relationship between risk perception and complacency   The underestimation of risk can lead to a reduced level of risk countermeasures, which in turn may lead to conditions of unsafety. Whether this process is the same as the onset of complacency is a question worth deeper investigation. Complacency in safety issues may be the the result of a number of factors, but what is definite is that a complacent attitude can be highlighted as a latent condition within a number of organisational accidents. This may be at any level from the operators to those at management and director level.

Whether the process of being complacent is as a result of personality traits (i.e. can individuals be *complacency-prone*?) or whether it is the result of incomplete information, is an important discovery to make. Even if the latter were found to be predominant, it may be that this process works in two directions. In other words, whereas a full appreciation of the risks of a particular activity may minimise the level of fallible decisions taken through naïvety or ignorance, it may also mean that excessive risk countermeasures or safety margins are reduced. However, the latter situation is difficult to accurately predict, especially in the case of complex high technology systems such as aviation.

Performance Indicators

The development of indicators of system health   The move towards more systemic investigation of accidents and incidents has helped to develop a greater understanding of system safety and the multiple factors which need to be in place to cause an accident. As such, the investigative emphasis has shifted from post-mortem towards that of investigating incidents as a means of prevention. This has in turn led to the stage of attempting to highlight latent conditions well before they compromise the efficiency of a system.

However, such a task is easier said than done. Although models such as those produced by Reason allow organisational failures to be more fully understood in terms of multiple causal factors, it stops short of being a proactive tool for actively monitoring system safety health. The challenge is to develop a simple, inexpensive methodology to accomplish this task which is attractive to all levels of operator, not just those with an active safety culture.

In Australia, the need to make inroads into the area of safety performance indicators was highlighted by the findings of the House of Representatives Standing Committee on Transport and Communications Infrastructure's Investigation in Aviation Safety. An initial attempt by BASI, CASA and the Department of Transport fell short of producing active indicators, plotting as they did, general trends over a ten year time period. A longer term and fresher strategy appeared in the shape of the INDICATE program which has been successfully implemented by a number of regional airlines.

A valuable area of research could be the development of a comprehensive set of system safety indicators which could be assessed using a simple checklist which would be weighted similarly to the Flight Safety Foundation's CFIT Checklist. Using the experience of complex high technology industries outside aviation, such as petrochemical and nuclear, a self-diagnosis package could be usefully made available to the aviation industry. The design must be simple, easy to use and need minimal training before implementation if it is to catch the full spectrum of operators and not just those with particularly safety-conscious attitudes.

There is much to be done and with traffic continuing to grow, maintaining the status quo is not an option. Aircraft such as the A380, carrying upwards of 600 passengers will soon be a reality and there are known problems that need solving. This case study has hopefully inspired some ideas for research and action, but the road ahead is a long one.

13 Epilogue

QF1: The Beginning of the End?

On 23rd September 1999, Qantas' impeccable safety record very nearly came to an end. A Boeing 747-438 aircraft carrying 391 passengers and 19 crew overran Bangkok runway 21L in a rainstorm. Coming to rest on a golf course 200m from the end of the runway, the aircraft was substantially damaged. Thankfully, there were no injuries to either crew or passengers.

The reaction within Australia was of shock for some and 'told you so' for others. Many who had long believed that an accident was inevitable seemed to be smugly smiling at their own apparent wisdom. The media was in a frenzy and no-one quite knew what the facts really were. Was it an accident or was it an incident? Was this Qantas' first accident ever? Was the aircraft a hull loss?

The facts are simple: It was an accident and not an incident. ICAO define an accident as involving major damage *or* a hull loss. It was not Qantas' first accident; there have been accidents involving Qantas jet aircraft and fatal accidents involving the propellor fleet. It was not however a hull-loss accident as the aircraft was repaired, albeit at considerable cost, and is back flying with the Qantas fleet.

The investigation, conducted by the Australian Transport Safety Bureau and Thai Authorities, has been extensive and has already resulted in a number of changes within both Qantas and the Civil Aviation Safety Authority. A final report is due to be published in 2001.

The important question for this book is whether this accident contradicts the research findings or makes them obsolete?

It's not a Competition

Aviation is a highly competitive business, but in safety, there is no competition. An accident does not render all of the information collected to date invalid or indeed necessarily signify a fall from grace. Qantas is still a safe airline. Indeed, just as Ansett did so successfully following its 1996 accident at Sydney, so Qantas has the opportunity to be a safer airline by learning from its mistakes.

The reasons behind Australia's good record for airline safety are no less valid now than before the accident. Whether all of the safety margins that were in place have remained so is an issue not just for Qantas, but for the whole aviation system. Errors which may have occurred on the day of the accident may well prove to be symptomatic rather than causal.

It is foolish to speculate on the contributory factors behind the Bangkok accident or indeed any accident before the investigation report is completed. However, it has become apparent that various changes have been taking place within the aviation system that present new hazards.

The Australian aviation scene continues to evolve as traffic grows and new demands are placed on the system. Domestic aviation exists within a global setting of economic cycles, regulatory changes technological advances and alliances. The last couple of years has seen Qantas become a member of the Oneworld airline alliance with airlines such as British Airways, American Airlines, Cathay Pacific and Aer Lingus. Ansett Australia meanwhile became a member of the Star Alliance following the initial acquisition of 50% of its shares by Air New Zealand. Subsequently Air New Zealand increased ownership to 100% and was then in turn bought into by Singapore Airlines. At the time of going to press, Ansett Australia and Air New Zealand were integrating their domestic and regional operations whilst Ansett New Zealand had become Qantas New Zealand.

Changes within the mainline carriers have been accompanied by growth in the regional carriers and the entry of two new high-capacity airlines. Qantas regional partners Airlink and Southern now operate BAe146 aircraft and Ansett regional, Kendell Airlines, has recently introduced 50 seat Canadair Regional Jets to its hitherto propellor-only operations. These bring a range of new challenges for crew training, maintenance, audit and so on, not least as the regional carriers tend to carry the tail livery of the mainline carrier, even when they may only be franchised operations.

New entrants, Virgin Blue and Impulse Airlines have started to operate high capacity jet aircraft on the main routes during 2000. Operating the B737 and B717 aircraft respectively, the carriers represent the new breed of budget carriers that have previously appeared in Europe and North America. Whilst no-frills carriers such as Southwest Airlines, Ryanair and Easy Jet have been extremely successful, many have also failed or been involved in accidents. (eg. Valujet)

Fares as low as $33 single Sydney-Melbourne were unprecedented in Australia and both Qantas and Ansett were forced to hit back to maintain market share. Low fares do not necessarily equate to poor safety, but prolonged aggressive competition has the capacity to weaken even established carriers. CASA recognise that poor financial health has been a contributory factor in a number of high profile accidents over the past few years.

The Australian aviation industry needs to be extremely careful not to trim back too many of the margins of safety without recognising the effect this can have on overall safety levels. Aiming only for regulatory compliance may be legally safe, but is unlikely to yield the same level of overall safety that exceedance of regulations brings.

Changes are occurring throughout the Australian system that are reducing safety margins. Arguably these are balanced out by improvements such as the mandating of TCAS for high capacity operators, implementation of TAAATS and so on. However, the question remains: Is aviation suffering from systemic homeostasis – the process whereby we have reached an acceptable level of safety and are now just looking for ways to do the same for less?

Only time will tell what the effect of change will be. However, as an industry, can we afford to just wait and see?

References

ABS - Australian Bureau of Statistics (1993) *Overseas Arrivals and Departures.* Australian Government Publishing Service, Canberra.

Air Safety Regulation Review Task Force (1990) *Air Safety Regulation Issues.* Second Report of the Aviation Safety Regulation Review. Australian Commonwealth Government Publishing, Canberra.

Albon, R. P. and Kirby, M. G. (1983) *Cost-padding in Profit-regulated Firms.* Economic Record 59 (164, March) p.16-27.

Alexander, K. and Stead, G. (1993) Aptitude Assessment in Pilot Selection. In R. A. Telfer (Ed.), *Aviation Instruction and Training.* Ashgate, Aldershot.

Ashford, N. J., Ndoh N. and A. S. Brooke, (1995) Airport Ramp Accident Study cited in Kilbride, D. A. (1996) *Ground Handling Claims - Who Pays? Does it Matter? And to Whom?* Proceedings of the 3rd International Aviation Conference, London.

Ashford, R. (1994) *Safety in the 21st Century - The Need for Focused Regulatory Targets and Maximised Safety Benefits.* Paper presented at the 47th International Air Safety Seminar, Flight Safety Foundation, Lisbon, Portugal, October 31st - November 3rd.

Baker, R. L. A. (1988) *Pilot Initial Intake Endorsement and Line Training.* Proceedings of the 41st Annual International Air Safety Seminar 'Basic Principles - the Key to Safety in the Future'. Flight Safety Foundation. December 5-8th, Sydney.

Baker, R. L. A. and Frost, K. (1993) Australian Airlines' Integrated Crew Training. In B.J. Hayward & A.R. Lowe (Eds.), *Towards 2000: Proceedings of the 1992 Australian Aviation Psychology Symposium.* Melbourne: The Australian Aviation Psychology Association.

Barlay, S. (1990) *The Final Call.* Arrow Books, London.

Barnett, A., Abraham, M and V. Schimmel. (1979) Airline Safety: Some Empirical Findings. *Management Science*, Volume 25, Number 11, November, pp.1045-1056.

BASI - Bureau of Air Safety Investigation (1991) *Limitations of the See and Avoid Concept.* Commonwealth of Australia, Department of Transport and Communications. Canberra.

BASI - Bureau of Air Safety Investigation (1994) *Investigation Report: British Aerospace BAe 146-200A VH-JJP, near Meekatharra Western Australia, 22 March 1992.* BASI, Canberra.

BASI - Bureau of Air Safety Investigation (1994b) *Investigation Report 9301743 - Piper PA31-350 Chieftain VH-NDU, Young, NSW. 11th June 1993.* Department of Transport. Canberra

BASI - Bureau of Air Safety Investigation (1995) *Air Safety Occurrence Report: 9501346.* Commonwealth of Australia, Department of Transport and Regional Development. Canberra.

BASI - Bureau of Air Safety Investigation (1996) *Investigation Report 9402804- Rockwell Commander 690B VH-SVQ en route Williamstown, NSW to Lord Howe Island,* 2nd October 1994. Department of Transport. Canberra.

BASI - Bureau of Air Safety Investigation (1996b) *Investigation Report 9403038 Boeing 747-312 VH-INH Sydney (Kingsford-Smith) Airport, New South Wales.* 19th October 1994. Department of Transport. Canberra.

BASI - Bureau of Air Safety Investigation (1996c) *Proactively Monitoring Airline Safety Performance - INDICATE.* BASI, Canberra.

Beaty, D. (1995) *The Naked Pilot   The Human Factor in Aircraft Accidents.* Airlife, Shrewsbury.

Bent, J. (1997) *Training For New Technology - Practical Perspectives.* Paper presented to the Ninth International Symposium on Aviation Psychology. April 27th - May 1st. Columbus.

BIE - Bureau of Industry Economics (1994) *International Performance Indicators - Aviation.* Research Report 59. Australian Government Publishing Service, Canberra.

Bigsworth, W. R. (1988) *An Overview of Air Traffic Control in Australia.* Paper presented to the 41st Annual International Air Safety Seminar of the Flight Safety Foundation, December 5-8, Sydney.

Blainey, G. (1977) *The Tyranny of Distance - how distance shaped Australia's history.* Sun Books, Melbourne.

Blomberg, R. D., Schwartz, A. L., Speyer, J-J., and J-P. Fouillot, (1988) *Application of the Airbus Workload Model to the Study of Errors and Automation.* Academie Nationale De L'Air Et L'Espace 3rd International Symposium on Aviation and Space Safety. 20-22nd September. Toulouse.

Boeing Airplane Safety Engineering (1993) *Statistical Summary of Commercial Jet Aircraft Accidents: Worldwide Commercial Jet Fleet 1959-1992.* Boeing Commercial Airplane Group, Seattle.

Boeing Airplane Safety Engineering (1996) *Statistical Summary of Commercial Jet Aircraft Accidents: Worldwide Commercial Jet Fleet 1959-1995.* Boeing Commercial Airplane Group, Seattle.

Boeing Commercial Airplane Group (1993) *Accident Prevention Strategies: Removing Links in the Accident Chain.* Boeing Airplane Safety Engineering, Seattle.

Braby, C., Muir, H. and D. Harris (1991) *The Development of a Working Model of Flight Crew Underload. Reproduced in Stress and Error in Aviation,* Proceedings of the XVIII Western European Association for Aviation Psychology. Avebury Publishing, Aldershot.

Bracalente, E. M. (1995) Aero-Space Technologist, Sensors Research Branch, National Aeronautics and Space Administration Langley Research Centre. Personal Correspondence.

Braithwaite, G. R. (1997) Get The Message! - Education is the Smart Side of Research. In Wiggins, M., Henley, I. and P. Anderson (eds.) *Aviation Education Beyond 2000.* University of Western Sydney, Macarthur.

BTCE - Bureau of Transport and Communications Economics (1992) *Social Cost of Transport Accidents in Australia.* Report 79. AGPS, Canberra.

BTR - Bureau of Tourism Research (1993) *International Visitor Survey.* Australian Government Publishing Service, Canberra.

Buddee, P. (1978) *Airways - The Call Of The Sky.* Lothian Publishing, Melbourne.

Burchill, K. (1994) Editor, IAPA World, Personal Correspondence, February 1994.

CAA - Australian Civil Aviation Authority (1994) *Airspace - Aspects of Design.* CAA Directorate Information and Communications Branch, 1994.

CAA - Australian Civil Aviation Authority (1994b) *Air Traffic Services Background Paper.* ACAA, September 1994.

CAA - Australian Civil Aviation Authority (ATS Division) (1994c) *Human Factors and Air Traffic Control; Some Thoughts for Controllers in the lead-up to TAAATS.* Paper presented at the 8th Biennial Convention of the Civil Air Operations Officers Association of Australia, Coffs Harbour, 9-12th November.

CAA - Civil Aviation Authority of Australia (1990) *Air Safety Regulation Review - Air Safety Regulation Issues.* Second Report of the Air Safety Regulation Review Task Force. Canberra: Australian Government Publishing Service.

CAA - UK Civil Aviation Authority (1982) *Human Factors In Accidents Due To Controlled Flight Into the Ground.* CAA, Gatwick.

CAA Board Safety Committee. (1993) *Report - Safety Regulation and Standards.* (Alan Terrell, Chairman), Canberra, Unpublished.

Caeser, H. (1994) Director, Aviation Safety Consulting (Former Lufthansa Chief Safety Pilot). Personal Correspondence, February.

Caine, J. (1994) Manager, Federal Airports Corporation, Perth Airport, Personal Correspondence, February.

Chalk, A. (1987) *Air Travel: Safety Through the Market.* International Society of Air Safety Investigators Forum, Washington DC, November.

Charlwood, D. (1981) *Take-Off To Touchdown - The Story of Air Traffic Control.* Australian Government Publishing Service, Canberra.

Chidester, T. and Vaughn, L. (1994). *Pilot/flight attendant coordination.* The CRM Advocate, 94(1), 8-10.

Chute, R.D. and Wiener, E.L. (1996). Cockpit/cabin communication: II. Shall we tell the pilots? *International Journal of Aviation Psychology,* 6(3), 211-231.

Collins English Dictionary (1993) Third Edition, Harper Collins Publishers.

Commonwealth of Australia (1997) *Aviation Safety Indicators - A Report of System Safety Indicators Relating to the Australian Aviation Industry.* 1997 Edition. Australian Government Publishing Service, Canberra.

Cooper, G.E., White, M.D., & Lauber, J.K. (1980) *Resource Management on the Flightdeck:* Proceedings of a NASA/Industry Workshop. (NASA CP-2120). Moffett Field, CA. NASA Ames Research Center.

Crowder, B. (1995) *Wonders of the Weather.* Australian Bureau of Meteorology, AGPS, Canberra.

CSS (Council for Science and Society). (1977) *The Acceptability of Risks.* Barry Rose Publishers, Ringwood.

Curtis, T. (1996) *Airline Accidents and Media Bias: New York Times 1978-1994.* Published on the World Wide Web.

Cushing, S. (1994) *Fatal Words: Communication Clashes and Aircraft Crashes.* University of Chicago Press, Chicago.

Davenport, J. K. (1988) *Qantas Pilot Employment and Indoctrination.* Proceedings of the 41st Annual International Air Safety Seminar 'Basic Principles - the Key to Safety in the Future. Flight Safety Foundation. December 5-8th, Sydney.

Department of Transport (1979) *Domestic Air Transport Policy Review.* Australian Government Publishing Service, Canberra.

Det Norske Veritas, (1979) *Safety, Man and Society.* Det Norske Veritas, Oslo.

DFS - Royal Australian Air Force Directorate of Flying Safety. (1994) *Review of the Boeing 707 Accident near RAAF Base East Sale on 29th October 1991*. Directorate of Flying Safety, Canberra.

Diehl, A. E. (1991) Cockpit Decision Making. *FAA Aviation Safety Journal*, Autumn, Washington, DC.

Doss, W. W. (1990) *Whoop Whoop Pull Up Pull Up!* International Federation of Airline Pilots Associations. Reprinted in Flight Deck. British Airways, Autumn 1992.

Dunn, M. D. (1988) *Aircraft Airworthiness: The Australian View*. Proceedings of the 41st Annual International Air Safety Seminar 'Basic Principles - the Key to Safety in the Future'. Flight Safety Foundation. December 5-8th, Sydney.

Dyster, B. and D. Meredith. (1990) *Australia in the International Economy in the Twentieth Century*. Cambridge University Press, Cambridge.

Edmonds, L. (1994) Capital - The Cause of Australia's First Airline Accident. *The Journal of Transport History*. Third Series. Vol. 15, No. 2. Manchester University Press. September.

Edwards, A. (1993) *Flights To Hell*, Thomas & Lochar, Nairn.

Englander, T., Farago, K., Slovic, P. and B. Fischhoff (1986) A Comparative Analysis of Risk Perception in Hungary and the United States. *Social Behaviour* 1 pp.55-66.

FAA - Federal Aviation Authority (1988) Advisory Circular AC00-54, *Pilot Windshear Guide*. US Department of Transportation.

Faulkner, J. P. E. (1996) Adjunct Associate Professor, Department of Aviation, University of New South Wales, Sydney. Personal Correspondence, October.

Felton, C. E. (1988) *A Pilot's View of Cockpit-Passenger Relations*. Paper presented at the Flight Safety Foundation 41st Annual International Air Safety Seminar 'Basic Principles - The Key to Safety in the Future'. December 5-8, Sydney.

Findlay, C. C., Kirby, M. G., Gallagher, F., Forsyth, P. J., Starkie, D., Starrs, M, and C. A. Gannon (1984) *Changes in the Air? - Issues in Domestic Aviation Policy*. Centre for Independent Studies, Australia.

Fire International (1993) Passengers Are in Danger at Small Airports - article by Stephen Lamble. Issue 138, February.

Flight International (1991) Bush-Bashing Experience - letter to the editor from Capt. P. Tomkinson, Germany. (pp.45) 16-22nd October.

Flight International (1994) Cultivating Safety. - article by Paul Phelan. (pp.22-24) 24-30th August.

Flight International (1994a) Culture is not the only safety issue. Letters to the editor. (pp.57) 28th September - 4th October.

Flight International (1994b) Culture v safety: apply caution. Letters to the editor. (pp.51) 9-15th November.

Flight International (1996) IATA Chief Proposes Culture / Safety Relationship Study. - article by David Learmount. (pp.8). 24-30th April 1996.

Flight Safety Foundation (1995) *Commuter Airline Safety - Special Safety Report*. Flight Safety Digest, January. FSF, Stirling, VA.

Flynn, J., Slovic, P. and C. K. Mertz (1994) Gender, Race and Perception of Environmental Health Risks *Risk Analysis*, Vol. 14, No. 6.

Forsyth, P. J. and Hocking, R. D. (1980) *Economic Efficiency and the Regulation of Australian Transport*. Council for the Economic Development of Australia, Melbourne.

FSF - Flight Safety Foundation (1994) *CFIT Checklist 'The CFIT Threat - Evaluate the Risk and Take Action.* Flight Safety Foundation, Arlington.

FSF - Flight Safety Foundation (1996) *Airport Safety: A Study of Accidents and Available Approach-and-landing Aids.* Special FSF Safety Report, Flight Safety Digest, March. Flight Safety Foundation, Arlington.

Goszczynska, M., Tyszka T. and Slovic, P. (1991) Risk Perception in Poland: A Comparison with Three Other Countries. *Journal of Behavioural Decision Making* 4, pp.179-193.

Graeber, C., Dement, W., Nicholson, A. N., Sasaki, M. and H. M. Wegmann (1986) International Cooperative of Aircrew Layover Sleep: Operational Summary. *Aviation, Space and Environmental Medicine*, 57 (12) pp. 3-19.

Graham, W. and Orr, R. H. (1970) *Separation of Air Traffic by Visual Means: An Estimate of the Effectiveness of the See and Avoid Doctrine.* Proceedings of the IEEE, Volume 58.

Green, R. (1993) Assistant Director, Army Personnel Research Establishment, Ministry of Defence. Personal Correspondence. 1st November.

Grieve, S. (1994) Vice President, Operations. Britannia Airways Ltd. Personal Correspondence. November 15th.

Gudykunst, W. B. and S. Ting-Toomey (1988) *Culture and Interpersonal Communication.* California, Sage Publications.

Gunn, J. (1988) *High Corridors - Qantas 1954-1970.* University of Queensland Press, St. Lucia.

Hall, R. (1994) Manager, Federal Airports Corporation, Brisbane Airport, Personal Correspondence. March.

Härtel, C. E. J.,and Härtel, G. F. (1995) *Controller Resource Management   What Can We Learn From Aircrews?* DOT/FAA/AM-95/21 Office of Aviation Medicine, Washington DC.

Hayward, B. J. and Lowe, A. R. (Eds.) (1993) *Towards 2000 Future Directions and New Solutions.* Proceedings of the 1992 Australian Aviation Psychology Symposium, Manly, 16-20th November. AAvPA.

Hayward, B.J. (1993) *The Dryden accident: A Crew Resource Training Video.* In R.S. Jensen & D. Neumeister, (Eds.), Proceedings of the Seventh International Symposium on Aviation Psychology. Columbus, OH: Ohio State University.

Hayward, B.J. (1995) Extending crew resource management: an overview. In N. McDonald, N. Johnston, & R. Fuller (Eds.), *Applications of psychology to the aviation system.* Aldershot, UK: Avebury Aviation.

Hayward, B.J. (1997) Human factors: Training for organisational change. In R.A. Telfer & P.J. Moore (Eds.), *Aviation Training: Learners, Instruction and Organisation.* Ashgate, Aldershot.

Hayward, B.J. (1997b) *Culture, CRM and Aviation Safety.* Paper presented to the Australian and New Zealand Societies of Air Safety Investigators Asia Pacific Regional Seminar 'Aviation Safety for the 21st Century in the Asia Pacific Region'. 30-31st May, Brisbane.

Helmreich, R. L. (1991) *Strategies for the Study of Flightcrew Behaviour.* Proceedings of the Sixth International Symposium on Aviation Psychology. (pp. 338-343) April 29th - May 2nd. Columbus.

Helmreich, R. L. and Wilhelm, J. A. (1997) *CRM and Culture: National, Professional, Organisational, Safety.* Paper presented at the Ninth International Symposium on Aviation psychology. April 27th - May 1st. Columbus, Ohio.

Helmreich, R.L. and Foushee, H.C. (1993). Why crew resource management? Empirical and theoretical bases of human factors training in aviation. In E. Wiener, B. Kanki, & R. Helmreich (Eds.), *Cockpit Resource Management*, (pp. 3-45). San Diego, CA: Academic Press.

Herald Sun (1995) Air Safety Bombshell. Article. 24th February, Melbourne.

Hewes, V. and Wright, P. H. (1992) Airport Safety. in Ashford, N. J and Wright, P. H (eds.) *Airport Engineering*. Third edition. John Wiley and Sons, New York.

Hines, W. and Helmreich, R. L. (1997) *Line Performance in Standard and Advanced Technology Aircraft*. Paper presented at the Ninth International Symposium on Aviation psychology. April 27th - May 1st. Columbus, Ohio.

Hofstede, G. (1980) *Culture's Consequences; International Differences in Work-Related Values*. Sage Publications, Beverley Hills, California.

Hofstede, G. (1991) *Cultures and Organisations: Software of the Mind*. McGraw-Hill, Maidenhead.

HORSCOTCI - House of Representatives Standing Committee on Transport, Communications and Infrastructure. (1995) *Plane Safe - Inquiry into Aviation Safety: The Commuter and General Aviation Sectors*. Australian Government Publishing Service, Canberra.

Hughes, D. (1994) Safety Group Highlights CFIT Risk for Regionals - Special Report, *Aviation Week & Space Technology*. 9th May.

IAPA - International Airline Passengers Association (1993) *Honour Roll of The World's Safest Airlines*. IAPA Travel Safety Alert, London. 22nd September.

IAPA - International Airline Passengers Association (1993b) The World's Safest Airlines. IAPA World Magazine, London. 22nd November.

IATA - International Air Transport Association (1993) *IATA Safety Record 1992*. IATA & Airclaims Ltd., Montreal.

ICAO - International Civil Aviation Authority (1990) *Annex 14 of the Chicago Convention*. ICAO Publications, Montreal.

ICARUS - Flight Safety Foundation Working Committee on Aviation Safety (1994) Flight Safety Digest, FSF, Arlington. December.

ISASI - The International Society of Air Safety Investigators (1983) *Code of Ethics*. Members Handbook. ISASI, Sterling.

Jensen, R. S. (1982) Pilot Judgment: Training and Evaluation in Hurst, R., and Hurst, L., *Pilot Error*. Granada Publishing, London.

Jensen, T. G. (1988) *Training of Technical and Cabin Crews for Emergencies*. Proceedings of the 41st Annual International Air Safety Seminar 'Basic Principles - the Key to Safety in the Future'. Flight Safety Foundation. 5-8th December, Sydney.

Job, M. (1991) *Air Crash - The Story of How Australia's Airways Were Made Safe*. Volume 1. Aerospace Publications Pty Ltd. Weston Creek.

Job, M. (1992) *Air Crash - The Story of How Australia's Airways Were Made Safe*. Volume 2. Aerospace Publications Pty Ltd. Weston Creek.

Johnson, B.,B. and V. T. Covello (eds.) (1987) *The Social and Cultural Construction of Risk*, 3-4. Reidel Publishing, Netherlands.

Johnston, N. (1993) Regional and Cross-Cultural Aspects of CRM. In Hayward, B. J. and Lowe, A. R. *Towards 2000 - Future Directions and New Solutions*. Proceedings of the 1992 Australian Aviation Psychology Symposium. Sydney.

Jones-Lee, M. W. (1989) *The Economics of Safety and Physical Risk.* Basil Blackwell, Oxford.

Karpowicz-Lazreg, C. and Mullet, E. (1993) Societal Risks as Seen by a French Public. *Risk Analysis*, 13 pp.253-258.

Kasperson, R. E. and J. X. Kasperson, (1984) *Determining The Acceptability of Risk: Ethical and Policy Issues.* A reprint from the Clark University Hazard Assessment Group Centre for Technology, Environment and Development, Clark University, Worcester.

Keith, L. A. (1997) *The Challenges Facing Aviation Safety Regulators.* Paper presented at the Royal Aeronautical Society 1997 Sir Charles Kingsford Smith Memorial Lecture, Sydney. 24th September.

Kenton-Page, J. (1995) *CRM Issues for Foreign Nationals.* Paper presented to the Ninth Royal Aeronautical Society Human Factors Group Miniconference 'Flight Deck and Organisational Cultures'. CAA Aviation House, Gatwick. October 12th.

Kilbride, D. A. (1996) *Ground Handling Claims - Who Pays? Does it Matter? And to Whom?* Proceedings of the 3rd International Aviation Conference, London.

Kirby, M. G. (1979) An Economic Assessment of Australia's Two-Airline Policy. Australian *Journal of Management* 4 (2, October) pp.105-118.

Kone, O. and Mullet, E. (1994) Societal Risk Perception and Media Coverage. *Risk Analysis*, Vol. 14, No. 1.

Lane, J. C. (1964) *The Money Value of a Man.* Paper presented to the ANZAAS 37th Congress, 20th January. Canberra.

Lane, J. C. (1973) *Human Factors in Transportation.* Paper presented to the 10th Annual Conference of the Ergonomics Society of Australia and New Zealand, November. Sydney.

Lane, J. C. (1994) Evidence Presented to the Royal Aeronautical Society (Australian Division) Investigation into General Aviation Safety. Monash University Accident Research Centre, Melbourne.

Lautman, L. G. and Gallimore, P. L. (1987) *Control of Crew-Caused Accidents.* Proceedings of the 40th Annual International Air Safety Seminar 'Human Factors and Risk Management in Advanced Technology'. Flight Safety Foundation. 26-29th October, Tokyo.

Learmount, D. (1987) Qantas - Safety and Monopoly. Flight International. Reed Publishing. London. pp.21-24. 5th December.

Learmount, D. (1993) Safety Editor, Flight International. Personal Correspondence, 5th November.

Lewis, K. S. (1993) Head of Safety and Environment, Qantas Airways Ltd. Personal Correspondence. 29th June.

Lewis, K. S. (1995) Head of Safety and Environment, Qantas Airways Ltd. Personal Correspondence, 8th January.

Lewis, K. S. (1997) Qantas Exceedence of Regulations Table. Company Document. October.

Mahon, P. (1981) *Report of the Royal Commission to Inquire into The Crash on Mount Erebus, Antarctica of a DC10 Aircraft.* Commonwealth Government. Wellington.

Male, C. (1997) Gaining the Safety Advantage. *Aerospace International.* Royal Aeronatical Society, London.

Marthinsen, H. F. (1989) *Another Look at the See and Avoid Concept.* ISASI Forum, December.

Maurino, D.E. (1997) *CRM: Past, Presnt and Future.* Proceedings of the Ninth International Symposium on Aviation Psychology. Columbus.

Maurino, D.E., Reason, J., Johnston, N. and Lee, R.B. (1995). *Beyond Aviation Human Factors.* Avebury Aviation, Aldershot.

Mayes, P. (1997) Manager, Safety Analysis Branch, Bureau of Air Safety Investigation. Personal Correspondence, 14th October.

McCarthy, J. and Serafin, R. (1984) *The Microburst Hazard to Aircraft from JAWS (Joint Airport Weather Studies Project)* Interim Report for 3rd Year's Effort (FY-84), National Centre for Atmospheric Research, Boulder, Colorado.

Mechitov, A. and Rebrick, S. B. (1990) Studies of Risk and Safety Perception in the USSR. In Borcheding, K., Larichev, O. I. and Messick, M. (eds) *Contemporary Issues in Decision Making.* Elsevier, Amsterdam.

Melvin W. W. (1994) *Windshear Revisited.* Condensed version of paper presented to the Society of Automotive Engineers (SAE Paper 901995) Air Line Pilot, November.

Merritt, A. C. (1993) *The Influence of National and Organisational Culture on Human Performance.* Paper presented at the Australian Aviation Psychology Association Industry Seminar, Sydney, 25th October.

Merritt, A. C. (1997) *Replicating Hofstede: A Study of Pilot Attitudes in 18 Countries.* Paper presented at the Ninth International Symposium on Aviation psychology. April 27th - May 1st. Columbus, Ohio.

Moshansky, V. P. (1992) *Commission of Inquiry into the Air Ontario Crash at Dryden, Ontario.* Minister of Supply and Services, Canada.

Moshansky, V.P. (1995). Foreword. In D.E. Maurino, J. Reason, N. Johnston, & R.B. Lee, *Beyond Aviation Human Factors.* Avebury Aviation, Aldershot.

Mulder, M. (1980) The Daily Power Game. cited in Hofstede., G., *Culture's Consequences.* (1980) Sage Publications, Beverley Hills, California.

National Geographic (Journal of the National Geographic Society) February 1988.

National Research Council (1983) *Low Altitude Windshear and its Hazard to Aviation.* National Academy Press, Washington.

National Transportation Safety Board (1992) *Special Investigation report: Flight Attendant Training and Performance during Emergency Situations.* (NTSB/SIR-92/02) Washington, DC.

Oster, C. V. Jr., Strong, J. S. and C. K. Zorn (1992) *Why Airplanes Crash -Aviation Safety in a Changing World.* Oxford University Press, New York. USA.

Patience, A. (1991) *Softening The Hard Culture.* pp.29-35 Mental Health In Australia, December.

Phelan, P. (1994) Safety in Numbers? Flight International. 14-20th December. Reed Publishing, London.

Pidgeon, N. F. (1991) Safety Culture and Risk Management in Organisations. *Journal of Cross-Cultural Psychology,* Vol 22 No.1, March. pp.129-140.

Potts, R. (1991) *Microburst Observations In Tropical Australia.* Paper presented at the Fourth International Conference on Aviation Weather Systems, June 24-26, Paris.

Proctor, P. (1993) Australia Reaps Benefits Of CAA Restructuring. *Aviation Week and Space Technology,* January 18th. McGraw Hill.

Qantas (1997) Qantas Exceedence of Regulations. Qantas Safety and Environment Department, October.

Quinn, M. (1997) Manager, Safety Investigation, Qantas Airways Ltd. Personal Correspondence.

Ratner Associates Inc. (1987) *Review of the Air Traffic Services System of Australia*, prepared for The Department of Aviation, Commonwealth of Australia, April.

Ratner Associates Inc. (1992) *1992 Review of the Australian Air Traffic Services System.* prepared for the Bureau of Air Safety Investigation, Commonwealth of Australia, April.

Reason, J. (1990) *Human Error*, Cambridge University Press, Cambridge.

Reason, J. (1992) Collective Mistakes In Aviation: "The Last Great Frontier", Article published in *Flight Deck* magazine volume 4 pp.28-34. British Airways, London.

Reason, J. (1993) Organisations, Corporate Culture and Risk. Paper reproduced in '*Human Factors in Aviation* - The Proceedings of the 22nd International Air Transport Association's Technical Conference'. IATA. Montreal, October.

Redding, S. G. and J. G. Ogilvie (1994) *Cultural Effects on Cockpit Communications in Civilian Aircraft.* Paper presented to the 37th Annual Flight Safety Foundation International Air Safety Seminar. Zurich. October 1994.

Reiner, A. (1992) Terminal Area GPWS Escape Considerations. Article published in *Flight Deck* magazine. British Airways, London.

Rohrmann, B. (1995) *Risk Perception Research   Review and Documentation* Programmgruppe Mensch, Unwelt, Technik (MUT) des Forschungszentrums Julich GmbH.

Royal Aeronautical Society Human Factors Group (1996) Discussions held at a miniconference held at the Civil Aviation Authority, Gatwick.

Royal Society (1992) *Risk: Analysis, Perception and Management.* The Royal Society, London.

Schiavo, M. and Chartrand, S. (1997) *Flying Blind, Flying Safe.* Avon Books, New York.

Sears, R. L. (1986) *A New Look at Accident Contributors and the Implications of Operational and Training Procedures.* Boeing Commercial Airplane Company, Seattle,.

Simkin, T. (1994) Volcanoes; Their Occurrence and Geography. from the proceedings of the First International Symposium on Volcanic Ash and Aviation Safety, published in *U.S. Geological Survey Bulletin* 2047.

Slovic, P., Fischhoff, B. and S. Lichtenstein (1980) Facts and Fears: Understanding Perceived Risk. Paper from '*Societal Risk Assessment*'. Proceedings of an International Symposium held October 8-9th 1979, GM Research laboratories, Warren, Michigan. Plenum Press, New York.

Smith, D. (1992) *Why Not A Safety Rating?* 33rd Sir Charles Kingsford Smith Lecture. Royal Aeronautical Society, Sydney.

Smith, D. (1993) Former Chairman of CAA. Personal Correspondence.

Smith, D. (1997) Keynote Speech. Risk Communication - The Second National Conference of the Risk Engineering Society. 2-3rd October, Canberra.

Smith, P. B., Dugan, S. and F. Trompenaars (1994) National Culture and the Values of Organisational Employees: A Dimensional Analysis Across 43 Nations. *Journal of Cross-Cultural Psychology* (in press).

Solomon, K. A. (1993) *Swimming Pool Risks: How Do They Compare to Other Accidental Risks?* RAND Report P-7841, Santa Monica.

Speyer, J-J. and Blomberg, R. D. (1989) *Workload and Automation. Human Error Avoidance Techniques:* Proceedings of the Second Conference. Aviation Research and Education

Foundation. 18-19th September, Virginia.

Spillane, K. I. and R. S. Lourensz (1986) *The Hazards of Horizontal Windshear to Aircraft Operations at Sydney Airport.* BMRC Research Report No.3. Melbourne.

Stead, G. (1995) Qantas Pilot Selection Procedures: Past to Present. In N. McDonald, N. Johnston, & R. Fuller (Eds.), *Aviation Psychology: Training and Selection.* Avebury Aviation, Aldershot.

Stewart, S. (1994) *Air Disasters.* Promotional Reprint Company Ltd. Leicester.

Stone, R. B. (1997) President's View. ISASI Forum - Journal of the International Society of Air Safety Investigators. April-June.

Svenson, O. (1979) *Are We All Among The Better Drivers?* Unpublished Report, Department of Psychology, University of Stockholm. Cited in Slovic, P., Fischhoff, B. and S. Lichtenstein (1980) Facts and Fears: Understanding Perceived Risk. Paper from 'Societal Risk Assessment'. Proceedings of an International Symposium held 8-9th October 1979, GM Research laboratories, Warren, Michigan. Plenum Press, New York.

Sydney Morning Herald (1994) Mumbo Jumbo: Ansett's Crash Riddle. 31st December.

TAC - Transport Accidents Commission (1995) *Comparison of Fatality Rates.* 5th June.

Taylor, L. (1988) *Air Travel - How Safe Is It?* BSP Professional Books, Oxford.

Technica Ltd. (1992) Cited in HSE Contract Research Report N0.33/1992 *Organisational, management and Human factors In Quantified Risk Assessment* Report 1. HMSO, London.

Teigen, K. H., Brun, W. and P. Slovic (1988) Societal Risk as Seen by a Norweigan Public. *Journal of Behavioural Decision Making.* 1, pp.111-130.

Terrell, A. (1995) Former Qantas General Manager, Flight Operations. Personal Correspondence, July.

Terrell, A. (1988) *Safety and Efficiency in International Airline Operation.* Aerospace, Royal Aeronautical Society, December.

Terrill, R. (1988) Australia at 200. *National Geographic* (Journal of the National Geographic Society), February.

The Age (1995) The Aviation Scandal. Editorial Opinion. Wednesday 19th April.

The Australian (1995) When Safety Comes Second. Feature article, p.27, 17-18th June, Melbourne.

Transportation Safety Board of Canada. (1995). *A Safety Study of Large Passenger-Carrying aircraft.* (SA9501). Ottawa: Minister of Supply and Services.

Trompenaars, F. (1993) *Riding the Waves of Culture - Understanding Cultural Diversity in Business.* The Economist Books, London.

Tylor, E. B. (1924) *Anthropology : an introduction to the study of man and civilization.* Macmillan Publishing, London.

Vail, G. J. and Ekman, L. G. (1986) Pilot-Error Accidents: Male vs Female. Ergonomics in Aviation. *Applied Ergonomics.* 17th December.4, pp.297-303, 17th December.

VRJ Risk Consultants (1994) Airspace Risk Analysis Generic Model. Company Literature, Melbourne.

Waldock, W. D. (1992) *Measuring Aviation Safety or The More Things Change, The More They Stay the Same.* ISASI Forum, March.

Ward, R. (1982) *Australia Since The Coming Of Man.* Lansdowne Press, Sydney.

Weber, J. (1994) Australian Representative on the Council of ICAO, Personal Correspondence, May.

Westrum, R. (1993) Cultures and Requisite Imagination. In Wise, J., Hopkin, V., D. and P. Stager (eds) *Verification and Validation of Complex Systems: Human Factors Issues.* Springer-Verlag, Berlin.

Westrum, R. (1995) Organisational Dynamics and Safety. In McDonald, N., Johnston and R. Fuller (eds) *Applications of Psychology to the Aviation System.* Avebury Aviation, Aldershot.

White, R. (1991) *Investing Australia: Images and Identity 1780-1980.* Allen and Unwin, Sydney.

Wolfe, T. (1979) *The Right Stuff.* Farrar, Straus, and Giroux, New York.

Woodhouse, R and R. A. Woodhouse (1997) *Statistical Measure of Safety.* Proceedings of the 9th International Symposium on Aviation Psychology, Columbus.

Wright, C. (1994) Manager, British Airways BASIS Incident Reporting System, Personal Correspondence, 11th August.

Index